CASCADIA'S FAULT

JERRY THOMPSON

CASCADIA'S FAULT

THE DEADLY EARTHQUAKE THAT WILL DEVASTATE NORTH AMERICA

HARPERCOLLINS PUBLISHERS LTD

To Bette and Ali, the people at the heart of my universe

Cascadia's Fault
Copyright © 2011, 2012 by Jerry Thompson.
Foreword © 2011 by Simon Winchester.
All rights reserved.

Published by HarperCollins Publishers Ltd

First published in Canada in a hardcover edition by HarperCollins Publishers Ltd: 2011
This HarperCollins trade paperback edition: 2012

HarperCollins books may be purchased for educational, business, or
sales promotional use through our Special Markets Department.

HarperCollins Publishers Ltd
2 Bloor Street East, 20th Floor
Toronto, Ontario
Canada M4W 1A8

www.harpercollins.ca

Library and Archives Canada Cataloguing in Publication
information is available upon request

ISBN 978-1-55468-467-0

Maps by Susan MacGregor / Digital Zone

Printed and bound in the United States
RRD 9 8 7 6 5 4 3 2 1

CONTENTS

CASCADIA SUBDUCTION ZONE

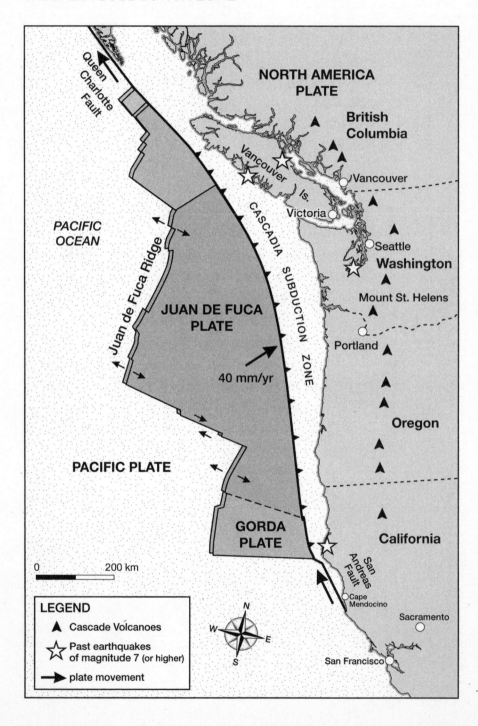

NORTH AMERICA PLATE

British Columbia

Vancouver

Victoria

PACIFIC OCEAN

Vancouver Is.

Queen Charlotte Fault

CASCADIA SUBDUCTION ZONE

Juan de Fuca Ridge

JUAN DE FUCA PLATE

40 mm/yr

PACIFIC PLATE

GORDA PLATE

Seattle

Washington

Mount St. Helens

Portland

Oregon

California

San Andreas Fault

Cape Mendocino

Sacramento

San Francisco

0   200 km

LEGEND

▲ Cascade Volcanoes

☆ Past earthquakes of magnitude 7 (or higher)

➔ plate movement

N W E S

TSUNAMI TRAVEL TIME

Each band represents an estimated one-hour increment of time for the tsunami to travel.

RING OF FIRE

LEGEND

- ▲ volcanoes
- ⌒ trenches
- ▨ Ring of Fire
- ☆ earthquake sites

1964 Alaska

Mt. Garibaldi
Mt. St. Helens
1700 Cascadia
1992 Cape Mendocino

1985 Mexico City

2010 Chile
1960 Chile

Puerto Rico Trench
Middle America Trench
Peru–Chile Trench
South Sandwich Trench

EQUATOR

Aleutian Trench
Kuril Trench
Japan Trench
2011 Tohoku
Izu Ogasawara Trench
Ryukyu Trench
Marianas Trench
Philippine Trench
Bougainville Trench
Tonga Trench
Kermadec Trench

2010 New Zealand

Mt. Pinatubo
Mt. Mayon
Krakatoa
2004 Sumatra
Sunda (Java) Trench

COMMUNITIES WITHIN RANGE OF CASCADIA'S FAULT

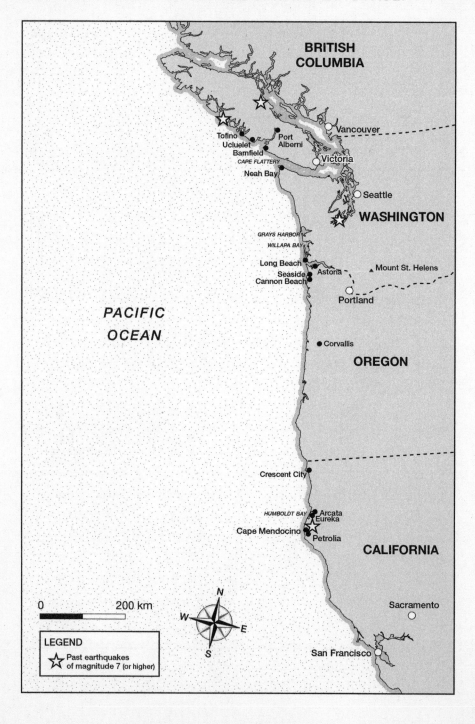

BRITISH
COLUMBIA

Vancouver

Tofino
Ucluelet
Port
Alberni
Bamfield
CAPE FLATTERY
Victoria
Neah Bay

Seattle

WASHINGTON

GRAYS HARBOR
WILLAPA BAY

Long Beach
Seaside
Astoria
Mount St. Helens
Cannon Beach

Portland

PACIFIC
OCEAN

Corvallis

OREGON

Crescent City

HUMBOLDT BAY
Arcata
Eureka
Cape Mendocino
Petrolia

CALIFORNIA

0   200 km

N
W   E
S

Sacramento

LEGEND
⭐ Past earthquakes
of magnitude 7 (or higher)

San Francisco

FOREWORD
by Simon Winchester

Of the sixteen most disastrous earthquakes to have shocked this planet since 1900, no fewer than fifteen have occurred along the shores of the Pacific Ocean, around the notorious Ring of Fire—a crucial but generally unfamiliar component of which is the subject of this book, the Cascadia Subduction Zone.

All of the earthquakes, and all of the tsunamis that some of these quakes have spawned, have been ferociously destructive. In drawing up a league table one might reasonably have supposed the Sumatran tsunami of December 2004—caused, of course, by a huge submarine earthquake after the fracture of an offshore subduction fault system— to be history's absolute worst; and insofar as it killed a quarter of a million people, then maybe in terms of statistical lethality it is. But what is now officially called the great Tohoku earthquake of 2011—and what spawned the grim tsunami that hit the northeastern Japanese coast on the afternoon of Friday, March 11—has implications that linger still, and that may well make this event even more deadly, in the long term.

The Japanese quake, originating as it did on the western edge of the Pacific tectonic plate, has revived interest—urgent and alarmed interest, even—in what might happen if the Cascadia fault system, which lies with ominous congruency on the plate's eastern side, were to rupture too. Most imagine the direst of consequences—consequences that are likely to parallel with some precision just what happened in Japan. The two fault systems are very similar and are tectonically connected to each other. Both lie roughly fifty miles (80 km) offshore from their respective continents; both are subject to vast internal stresses; both, if they fracture, can cause terrible ground-shaking onshore; and both can generate immense tsunamis at sea.

All of the broken bones that resulted in northern Japan, all of those broken lives and those broken homes and broken cities and, most sinister of all, those broken atomic reactors, swiftly prompted much of humankind to remember what in calmer times we prefer to forget: that most stern and chilling of mantras which holds, quite simply, that mankind inhabits this earth subject to geological consent—which can be withdrawn at any time.

For thousands of people, maybe tens or even hundreds of thousands, this consent was withdrawn with shocking suddenness (all geological events are sudden, and all are unexpected, if not necessarily entirely unanticipated) at 2:46 p.m. on what was a clear, cool spring afternoon. One moment all were going about their quotidian business—in offices, on trains, in rice fields, in stores, in schools, in jails, in warehouses, in shrines—and then the ground began to shake.

At first, the shock was merely a much stronger and rather longer version of the temblors to which most Japanese are well accustomed. There came a stunned silence, as there always does. But then, the difference: some few minutes later, a low rumble from the east, and in a horrifying replay of the Indian Ocean tragedy of just six years before, the imagery of which is still hauntingly replayed in all the world's minds, so the coastal waters off northern Honshu vanished, sucked mysteriously out to sea.

The rumbling continued as people began to spy a ragged white line on the horizon, and then, with unimaginable ferocity, the line became visible as a wall of waves sweeping back inshore at immense speed and at great height. Just seconds later, these Pacific Ocean waters hit the Japanese seawalls, surmounted them with careless ease, and began to claw across the land beyond in what would become a dispassionate and detached orgy of utter destruction.

We all now know, and have for fifty years, that geography is the ultimate reason behind these disasters. Japan is at the junction of a web of tectonic plate boundaries that makes it more peculiarly vulnerable to ground-shaking episodes than almost anywhere else—and it is a measure of Japanese engineering ingenuity, social cohesion, the ready acceptance of authority, and the imposition of necessary discipline that allows so many to survive these all-too-frequent displays of tectonic power.

But geography is not the only factor in this particular and acutely dreadful event. Topography played an especially tragic role in the story too—for it is an axiom known to all those who dwell by high-tsunami-risk coastlines that when the sea sucks back, you run—you run inland and, if at all possible, you run uphill. But in this corner of northeast Japan, with its wide plains of rice meadows and ideal factory sites and conveniently flat airport locations, there may well be a great deal of inland—but there is almost no uphill.

Such mountains as there are sit far away, blue and distant in the west. All here is coastal plain. And so the reality is this: if a monstrous wave is chasing you inland at the speed of a jetliner, and if the flat topography all around denies you any chance of sprinting to a hilltop to escape its wrath, then you can make no mistake: the wave will catch you, it will drown you, and its forces will pulverize you out of all recognition as a thing of utter insignificance—which of course, to a tsunami, all men and women and their creations necessarily must be.

Even more worrisome than geography and topography, though, is geological history. For this event cannot be viewed in isolation. There was a horrifically destructive Pacific earthquake in New Zealand on February 22, 2011, and an even more violent magnitude 8.8 event in Chile almost exactly a year before. All three phenomena involved more or less the same family of circum-Pacific fault lines and plate boundaries—and though there is still no hard scientific evidence to explain why, there is little doubt now that earthquakes do tend to occur in clusters: a major event on one side of a major tectonic plate is often (not invariably, but often enough to be noticeable) followed some weeks or months later by another on the plate's far side. It is as though the earth becomes like a great brass bell, which when struck by an enormous hammer-blow on one side, sets to vibrating and ringing all over.

And now there have been three catastrophic events at three corners of the Pacific plate—one in the southeast, on February 27, 2010; one in the southwest, on February 22, 2011; one in the northwest, on March 11, 2011.

That leaves just one corner unaffected—the northeast. And the fault lines in the northeast of the Pacific plate are most notoriously the San Andreas fault, underpinning the city of San Francisco; the Hayward fault, underlying the teeming and less well engineered cities of Oakland and Berkeley; and then offshore, the 800-mile-long Cascadia Subduction Zone fault. If the San Andreas or the Hayward fault were to rupture, the devastation on land could be immense; but if the Cascadia were to do likewise, the event would be global in scale, regional in destructiveness, lethal in a vast swath of countryside from Vancouver Island to the border of California. It would be a disaster of titanic proportions. It would, in a word, be epic.

That makes the geological community very apprehensive. All know that both the San Andreas fault and Hayward fault are due to rupture any day—the former last did so in 1906, and strains have built beneath it to a barely tolerable level. But it has been fully 311 years since Cascadia fractured—and it is in my view far from wholly irresponsible (though some scientists say otherwise) to employ the word *impending* when describing its potential to fracture and possibly cause a terrible disaster.

For Cascadia to rupture again, with unimaginable consequences for the millions who live in its danger zone, some triggering event has to occur. Three disasters—the Japanese tragedy of March 2011 being the most recent, New Zealand and Chile before that—have now taken place, and each of these can reasonably be regarded, perhaps, as a triggering event. There are in consequence a lot of thoughtful people in the American and Canadian west who are currently very nervous indeed—wondering, as they often must do, whether the geological consent that permits them to inhabit so pleasant a place might be about to be withdrawn, sooner than they had supposed.

INTRODUCTION

On Christmas Eve 2004 my wife, Bette, and I were in a hotel bar in San Francisco dreaming up plot points for a film we'd like to shoot some day when a woman arrived from the airport with breathless news. The bartender clicked his remote and *It's a Wonderful Life* vanished, Jimmy Stewart's smiling face wiped off the screen by a mountain of angry seawater. I can still see those endlessly repeated loops of amateur video shot from the balconies of beachfront resorts in Sumatra and Thailand, relayed by satellite to every TV receiver on the planet.

The first horrifying, mesmerizing wave crashed against a seawall, jetting skyward in salty white torrents, tearing through a fringe of palm trees like a monsoon river, across a hotel pool deck and a manicured square of green lawn. The darkening surge roared uphill through narrow, cluttered streets choked with tourist luggage, broken timbers, small motorcycles with their riders struggling to stay vertical, cargo vans overturned and bulldozed by white froth into market stalls. A transit bus floating on its side began to sink as desperate passengers jumped from the slippery roof.

It's impossible to forget the images, those flailing human bodies—especially one unfortunate older man clinging to the outside railing of a rapidly filling parkade. Exhausted and in shock, he finally let go. We watched as he sank into the muddy torrent and was swept away.

More than 230,000 people in fourteen countries around the Indian Ocean died or disappeared, many of them before our eyes, and there was nothing any of us could do. Everything not nailed to the ground was torn loose and carried off by the roaring water. And there was more to come. Even after the first water had cut a swath nearly a mile inland and then sucked itself halfway out to sea again, full of death and floating debris, people standing among the palms were so stunned by the spectacle they waited too long to outrun the next wave.

Most victims, including those who'd lived their entire lives along the beach—even fishermen who knew the sea quite well—had no idea these giant ripples would come ashore again and again. In Phuket, Thailand, some of the swells were sixty-five feet (20 m) high. Closer to the earthquake zone, in Aceh province on the northern end of the island of Sumatra, the mountain of water topped more than a hundred feet (30 m).

Until that moment, only a handful of people in the world had ever experienced a tsunami. Fewer still had any concept of what causes these so-called tidal waves. The magnitude 9.2 earthquake, generated by the movement of two tectonic plates along an almost nine-hundred-mile (1,400 km) undersea fault called the Sunda Trench, was never more than a footnote in the nonstop cycle of dismal news. The last time anything this big had happened in the Indian Ocean was more than six hundred years ago—so far back there were no written records, nor any social memory of the disaster. Perhaps that explains why so many were caught by surprise.

But the Indian Ocean disaster is only the most vivid example of what has happened before—and what lies ahead. Chile in 1960 had a magnitude 9.5 quake in which more than 2,000 lives were lost and 3,000 people were injured. Two million were left homeless. The resulting tsunami killed another 61 people in Hawaii, 138 more in Japan, and 32 in the Philippines. Alaska in 1964 suffered a magnitude 9.2 quake, with 128 lives lost and $311 million in property damage. Mexico City in 1985 was shaken by a magnitude 8.1 temblor in which at least 9,500 were killed, more than 100,000 were made homeless, and more than $3 billion of property damage was done. What happened to Sumatra in 2004 will also happen to California, Oregon, Washington, and British Columbia.

The geologic source of the looming catastrophe along North America's west coast—like all the others—lies hidden beneath the sea, out of sight and pretty much out of mind. Scientists, civil engineers, and emergency planners know with certainty that it's bound to happen here, but they're having a devil of a time getting anyone to pay attention. This book, I hope, will change that.

People in the United States and Canada, if they think at all about earthquake disasters, probably conjure up the infamous San Andreas fault as the worst case. In California, waiting for "the Big One," people wonder which city the San Andreas will wreck next—San Francisco or Los Angeles? Well, perhaps neither, because if by the Big One they mean the earthquake that will wreak havoc over the widest geographical area, that could destroy the most critical infrastructure, that could send a train of tsunamis across the Pacific causing economic mayhem that would probably last a decade or more—then the seismic demon to blame could not possibly be the San Andreas. It would have to be Cascadia's fault.

The Cascadia Subduction Zone is a crack in the earth's crust, roughly sixty miles (100 km) offshore and running eight hundred miles (1,300 km) from northern Vancouver Island to northern California. It has generated massive earthquakes not just once or twice, but over and over again throughout geologic time. A recently published, peer-reviewed scientific research paper documents at least forty-one

Cascadia "events" in the last ten thousand years. Nineteen of those events ripped the fault from end to end, a "full margin rupture."

As for timing, scientists used to think these mega-quakes occurred every 500 to 530 years, but the newest data show that the fault has at least four segments, the southernmost being far more active and with a greater number of slightly smaller (magnitude 8 or higher) quakes. Based on historical averages, the southern end of the fault—from Cape Mendocino, California, to Newport, Oregon—has a large earthquake every 240 years. For the northern end—from mid-Oregon to mid-Vancouver Island—the average "recurrence interval" is 480 years, according to a recent Canadian study. And while the north may have only half as many jolts, they tend to be full-size disasters in which the entire fault breaks from end to end at magnitude 9 or higher.

Given that the last big quake was more than 310 years ago, one might argue that a very bad day on the southern segment is ominously overdue. With a timeline of forty-one events an American geologist has now calculated that the California–Oregon end of Cascadia's fault has a 37 percent chance of producing a major earthquake in the next fifty years. The odds are 10 percent that an even larger quake will strike the upper end (in a full margin rupture) in fifty years. It appears that three centuries of silence along the fault (Cascadia is classified as the quietest subduction zone in the world) has been entirely misleading. The monster is only sleeping.

Cascadia is virtually identical to the offshore fault that devastated Sumatra—almost the same length, the same width, and with the same tectonic forces at work. This fault can and will generate the same kind of earthquake we saw off Sumatra: magnitude 9 or higher. It will send crippling shockwaves across a far wider area than all the California quakes you've ever heard about. Cascadia's fault will slam five cities at once: Vancouver, Victoria, Seattle, Portland, and Sacramento. It will cause physical damage as far south as San Francisco.

Cascadia's fault will cripple or destroy dozens of smaller towns and coastal villages from Tofino and Ucluelet on Vancouver Island to Crescent City and Eureka in northern California. None of these cities and towns will be able to call their neighbors for help because they will all be on their knees in rubble at exactly the same moment. California, with all its hard-earned earthquake experience, won't be able to offer much assistance to Oregon or Washington because it will have too many emergencies of its own to cope with. There will be no cavalry racing over the hill to save the day, no government white knights to bail anybody out. It'll be every man, woman, and child for themselves in three American states and a Canadian province.

The San Andreas has and will again cause terrible, destructive earthquakes, probably sooner rather than later, but the offshore temblor from Cascadia will be on a much larger scale. The San Andreas will wreck *a* major urban area—perhaps San Francisco *or* Los Angeles—but probably not both at the same time. Cascadia's fault will hammer an entire region of the planet, just as the Sumatra disaster did.

I know it's not scientific, but just for the sake of comparison, I fashioned a small ruler out of a piece of paper and held it up against a globe. I put a pencil mark midway up Vancouver Island and another at Eureka, California, creating a paper proxy for Cascadia's fault. Then I moved it to the east coast of North America. An eight-hundred-mile (1,300 km) tectonic crack like this—if it started in New York—would run down through Philadelphia, Baltimore, and Washington to about Charleston, South Carolina. The same fracture, if it started in Detroit, would run past Niagara Falls, through Toronto and Montreal to roughly Quebec City. Or from London through Paris to Nice, France. Or from Lisbon across Portugal to Barcelona. Or from Berlin to Milan.

There's evidence to suggest that Cascadia's fault would not act alone—that an earthquake along this subduction zone could transfer enough stress to *trigger* the San Andreas at the point where the two faults connect, near Cape Mendocino, about three hundred miles (480 km) north of San Francisco. Mud-core sampling of undersea landslides off the Pacific Northwest and California coasts has revealed a fairly close time correlation between ruptures on the two fault systems for many of the largest quakes in the past.

It could happen like this: a magnitude 9 rupture on Cascadia causes an unprecedented natural disaster that affects the entire Pacific Rim. It also sends a strong jolt into the neighboring San Andreas system, which is already nearing its own failure along the Hayward fault in Oakland. Ten or twenty years later, while Americans and Canadians are still rebuilding from the recent Cascadia event, the San Andreas rips loose and California is back on its knees again.

That's not all. Cascadia will also slam the beaches of the west coast of North America as well as Alaska and Hawaii. A research plan prepared by NOAA—the National Oceanic and Atmospheric Administration— back in 1982 estimated that 900,000 people would be at risk from a fifty-foot (15 m) Cascadia tsunami striking the U.S. western seaboard.

But that's just the United States. Nobody has done a projected death toll for the other Pacific Rim nations that would be affected. Researchers have, however, made a convincing case that an earthquake on Cascadia's fault in 1700 put a series of waves thirteen to sixteen feet (4–5 m) high—imagine water more than fifteen feet above the highest tides— onto the beaches of eastern Japan, causing widespread damage, injuries, and deaths. At this point one can only imagine what the same waves would do to the seaports and villages of modern-day Japan. To this scenario add Indonesia, Hong Kong, the Philippines, New Guinea, Australia, and New Zealand—all of which would be hit by Cascadia's waves.

And now we've learned that the effects could be even worse than previously imagined. The evidence from Sumatra, translated into numerical code and applied to updated computer models of Cascadia, confirms that some of the waves generated could be as high as seventy to ninety feet (20–30 m). Earlier computer simulations had put the maximum wave height at roughly fifty to sixty-five feet (15–20 m). Looking at footage of what a ninety-foot wall of water had done to one beach near Banda Aceh in 2004, an emergency manager from FEMA, the Federal Emergency Management Agency, shook his head and told a journalist from CNN that quite frankly no town on the western U.S. coast had any idea how to plan for or cope with that kind of wave.

Canada's situation will be considerably worse. Federal and provincial emergency planners in British Columbia are laughably underequipped and underfunded. Canada doesn't even have a national guard to take over when local governments are quickly overwhelmed—as they surely will be when Cascadia rips loose.

At NOAA's laboratory in Seattle, chief oceanographer Eddie Bernard says the easiest way to put Cascadia's fault into perspective is to "multiply the New Orleans disaster times four or five." Imagine having five Hurricane Katrinas—hitting five different cities—on the same day. Other experts believe Cascadia's next big temblor will be the largest peacetime disaster in North American history.

As a journalist and documentary filmmaker I have covered the Cascadia earthquake story at least five times over the past twenty years and have talked to dozens of geologists, seismologists, civil engineers, and oceanographers. With each new evolution or refinement of earthquake and tsunami science, I have been stumped in the final scene by the same nagging question. Do I make things better or worse by warning people about an event that may not happen in their lifetimes? If I take the latest scientific evidence, show that a monumental disaster is probably overdue and issue a dire warning—what do I accomplish? The initial shock wears off quickly. After that, nothing much changes.

All I've done in telling the tale is to make people more depressed than they already were. Folks who live around the Pacific Rim have heard this wake-up call many times before and disaster fatigue creates inertia. Because there are so many worrisome things to obsess about— global warming, terrorist attacks, killer plagues, and asteroid impacts, to name a few—disaster stories are losing their punch.

Unfortunately Cascadia's menace remains, whether those of us in the danger zone respond to it or not. What happened to Sumatra in 2004 will happen to North America, beyond any reasonable scientific doubt. A nearly identical earthquake will rattle the West Coast and a train of killer waves will tear across the Pacific.

One might argue that because I know what's going on, I have a moral obligation to spread the word even if I don't know how to respond to it myself. A replay of the Indian Ocean disaster along the shores of the North Pacific is absolutely going to happen. And Bette and I will be in harm's way when it does.

We live on a bluff above the sea on the mainland coast of British Columbia, about two hours' drive north of Vancouver. We have no idea how well our condo will withstand four minutes of intense shaking. Civil engineers generally agree that well-built wood-frame structures are flexible enough to ride out most earthquakes without collapsing, but Cascadia's shockwaves will reverberate far longer than most quakes. We tell ourselves we'll be okay, that the condo may be damaged but we'll probably get out alive. How then do we cope with the aftermath?

We're two hours away from the nearest big city, Vancouver, which may be in much worse shape than our own coastal hamlet. Rescue squads, medical aid, and emergency supplies will go first to the areas of greatest need, where the greatest numbers of people are affected—and where television cameras will focus the world's attention. To us this means help may be a long time coming to Sechelt, so we're pretty much on our own.

When they hear that I'm working on the Cascadia story yet again, friends and neighbors usually cringe. The first question everyone asks is when it's going to happen. I tell them I don't know, but some very smart people are trying to figure it out. Then they shrug. What can we do? It's too overwhelming to think about.

PART 1 TREMORS AND RIPPLES

CHAPTER 1

Mexico City: Preview of Coming Events

On the night of September 19, 1985, a jetliner packed with journalists, foreign rescue workers, and worried family members banked low across the flanks of Mexico City. The twinkling sprawl of suburbia gradually gave way to a black hole at the heart of the city. Only the twisting flames of unquenched fires and pockets of light powered by emergency generators penetrated the gloom. From window seats on final approach it was impossible to see the full extent of damage down below, but every person on that plane knew they were about to land in the haze of an ongoing nightmare.

Even on the ground visibility was so limited the arriving passengers could not appreciate how bad things were that first night. Robb Douglas, a Canadian television news cameraman, went to work straight away using a battery-powered lamp to shoot pictures of rescue teams digging through hunks of broken concrete and twisted steel. All he could capture in the small beam of light projected from his camera were close-cropped images of frenzied workers—their grim, sweat-streaked, exhausted faces. Dust and smoke and death.

Early the next morning the scale of the disaster revealed itself.

"When the sun came up you could see—holy shit! It looked like it had been bombed—like the city had been bombed," Robb would tell me later. "All those buildings had just dissolved." In little more than three minutes of horrific shaking, 10,000 people had died. Unofficial estimates would later push the death toll to 40,000 or more. Another 50,000 were injured and 250,000 were homeless.

In the city proper, 3,124 buildings were severely damaged and 412 had collapsed. If you included the impact on outlying regions, more than 6,000 buildings had either been destroyed or so heavily damaged they would have to be demolished. With a population of eighteen million, the world's second largest city was reeling from the blow and woefully unprepared for the aftermath.

Those of us who did not see it first-hand, who did not stand like Robb on the brink, who did not smell the blood or hear the screams barely paid attention. If we didn't know someone who lived there, it was one more disaster, far away and too gruesome to think about. But flash back several months to the summer of 1985 and consider the context.

On June 23, a terrorist's bomb blew Air India flight 182 from the sky off the Atlantic coast of Ireland, killing 329 passengers. On July 19, the Val di Stava dam in northern Italy collapsed, killing 268 people. On August 2, Delta Air Lines flight 191 crashed near Dallas, killing 137 people. Ten days later a Japan Airlines flight from Tokyo to Osaka crashed, killing 520, the deadliest single-aircraft accident in history. A crash on September 6 outside Milwaukee killed 31 more. Disasters had dominated the summer headlines with depressing regularity.

On Thursday, September 19, the morning the earthquake wrecked Mexico City, I was working in New York on a Canadian Broadcasting Corporation television documentary about a handful of Russian soldiers who had defected in Afghanistan and escaped to the United States. By the time the first shockwave (magnitude 8 on the Richter scale) hit the Mexican capital at 7:19 a.m. local time, the CBC camera crew and I had already left our Manhattan hotel. The front page of the

New York Times carried stories about an American hostage in Lebanon freed after sixteen months of captivity and a feature about police corruption in Philadelphia but nothing about Mexico City, of course, because the paper had gone to bed the night before.

Off to a good start on the Soviet defector story, I worked my way through the first full day of Mexico's tragedy without hearing a word about it. I was focused on "Russia's Vietnam," blissfully unaware of the mounting death toll in that smoky, flickering cauldron more than 1,800 miles (3,000 km) to the south and west. I did not know about heroic efforts underway to tunnel beneath the buckled walls and floors of a hospital to find survivors. I could not see the heartbreaking pictures my friend Robb was shooting at that very moment.

Even had I heard a radio newscast as we rattled over potholes and lurched back into New York just before midnight, the Mexico City story probably would not have fazed me as much as it should have. I had never experienced a disaster like that myself, so I had no effective way to process the information, no visceral sense of what it was like to feel the earth heave or to see the known world come crashing down around me. My awareness of distant tragedies was limited to a mental montage of anonymous, grieving widows, orphaned children crying, broken men with vacant stares, collapsed buildings, and body bags.

About a dozen years in television news and documentaries had sent me around the world more than once, but somehow I'd never been assigned to cover a natural disaster. An evolving cynic at the ripe old age of thirty-six, I had filed plenty of stories about human mayhem from places where the roads were mined and people tended to shoot at each other—places like Nicaragua and Honduras, Sri Lanka and the Punjab. But floods, typhoons, and earthquakes were *terra incognita*.

For me these doomsday stories had an air of unreality about them. The numbing repetition of so much tragedy, packaged neatly into two-minute doses and delivered almost instantly by television news, blunted my reaction. Disaster shock tended to fade quickly.

I was ignorant not just of the people struggling to survive that night in the Mexican capital but of the scientific significance of the earthquake itself. I would not learn until weeks later that the "event" (as geologists would refer to it) had come as a bit of a surprise. The offshore fault that ruptured had been quiet for perhaps two hundred years even though smaller earthquakes had occurred on either side of the rupture zone.

The quiescence of the Michoacán segment, roughly 190 miles (300 km) southwest of Mexico City, had convinced some seismologists that this particular fragment of the earth's crust was for reasons unknown a special case. Two tectonic plates were apparently sliding past each other smoothly. No stress, no worries.

Was there some kind of subterranean lubrication that kept the plates from getting jammed up? No one knew. But for as long as humans had kept written records, this zone and a few others like it around the world had not generated large, destructive earthquakes. Therefore, this particular stretch of the Mexican coast was thought to be aseismic—not likely to produce major earthquakes.

At sunset on Friday, September 20, Robb Douglas and his soundman Gunter Mende stood at the base of what might once have been a thirty-story apartment building but was now a twisted heap. Next to the destroyed building stood two more towers that looked to be part of the same complex. What caught Robb's eye was the apparent randomness of the wreckage. It was obvious the other two had been badly damaged and could collapse at any moment, but they were still vertical. Why? His pictures raised questions that civil engineers would be forced to answer in the coming months.

As Robb lined up his next shot, the ground started to shake violently beneath his feet. At 6:30 p.m., thirty-six hours after the initial shockwave, a magnitude 7.5 aftershock (some experts believe it was another, completely separate earthquake) instantly made a bad situation worse. People who'd counted themselves lucky, who'd thought they had survived the disaster, realized in a heartbeat that it wasn't over yet. Hundreds ran screaming from homes and apartments in the congested neighborhood surrounding the three towers.

They swarmed across sidewalks and into the streets. Some scrambled into cars and tried to make a getaway, racing toward a nearby freeway on-ramp. "I got shots of everybody running through the streets, trying to get in their cars to drive away," said Robb, "but it was hopeless." A torrent of humanity choked the only avenue of escape.

The TV camera was locked on a tripod and the recording deck was connected by a thick black cable, so Robb and Gunter were shackled together. With nowhere to run, they hunkered down, two rocks in a river of terrified people. "It happened so quickly. It's not like you have any warning." The more Robb told me the more vividly the scene came back into focus for him. "We were so frightened, we couldn't even move. I was just weak in the knees." This from a guy who doesn't get rattled by much.

Sensing movement over his shoulder, Robb glanced up and saw in a darkening sky that the two remaining towers, thirty stories high, were bending from side to side like tall trees in a wind storm. "When we saw those other two buildings moving back and forth like that, swaying—and you know one had already collapsed—and we were right beside 'em, we figured we were in the wrong place." But Robb and Gunter were lucky that night. The towers did not fall. And they lived to tell the tale.

Mexico had survived plenty of big earthquakes—forty-two of magnitude 7 or higher in the twentieth century—but nothing as big as the shockwave of September 19 had occurred there in all of recorded history. So geologists immediately started looking for ways to make sense of what had happened. An eerily similar offshore quake had rocked Alaska in 1964. Chile's 1960 seafloor rupture, at magnitude 9.5, was the largest earthquake ever recorded in the world. In both cases the faults that caused the quakes were impossible to examine, concealed beneath thousands of feet of seawater.

Geologists who'd suggested the Michoacán segment of the Mexican seaboard was somehow a special case—an aseismic zone—were about to face a steep new learning curve. But they weren't alone. Earth scientists around the world were undergoing a paradigm shift, a fundamental change in thinking about how the planet had formed, how mountains were built, and what makes huge earthquakes happen. For more than three decades, starting in the mid-1950s, much of the conventional wisdom of geology had been debated, updated, and revised by fresh data and new ideas.

By the mid-1960s a new hypothesis called plate tectonics had emerged from the dust of an earlier and much ridiculed theory known as continental drift (more on this later). But in 1985, twenty years after tectonic papers began to appear in the science literature, the idea was still relatively new and more than a little controversial. Not everyone had come to terms yet with the concept of slabs of the earth's crust, forty to sixty miles (70–100 km) thick, floating around on convection currents of superheated, semi-liquid rock, crashing and grinding against each other, creating jagged mountain ranges like the crumpled fenders of a car wreck, generating giant earthquakes in the process.

Even basic things like the textbook definition of a fault—a rupture in rock along which movement has taken place—had become vastly more complex in light of new discoveries. It turns out not all faults are simple fractures near the surface on dry land. Unlike the glaringly obvious San Andreas in California, where two plates are sliding past each other horizontally (where a geologist could easily stand with one foot on the North American plate and the other on the Pacific plate and straddle the fault to study it), these offshore rupture zones remained a mystery. Was the boundary between two plates always vertical? Or could one plate slip underneath another? And if so, at what angle? How could you prove it one way or the other? These were just some of the unknowns that would generate a spirited exchange in the coming years.

In the immediate aftermath of the Mexico City disaster, seismologists, marine geologists, and engineers tried to draw conclusions about the underlying cause and what it might mean for other supposedly aseismic zones around the world. Perhaps these monster quakes had happened before but researchers had not looked far enough into the past to find the evidence. Perhaps "all of recorded history" was simply too brief, in geologic terms, to see a repetition of these enormous undersea earthquakes. If it takes several centuries to build up enough stress for a quake this big, perhaps the last one happened so long ago there was nobody around to write it down.

Some scientists thought the Michoacán zone was a "seismic gap" where strong earthquakes (in 1939, in 1973, and again in 1979) had relieved stress on either side of the main segment, but the part in the middle—the Michoacán zone itself—had remained stuck for nearly two centuries. The zone was the only part of this tectonic plate that had not snapped free of the continental crust that had drifted over it. It was a holdout—a ninety-mile (145 km) slab of sub-sea rock that was bound to rip loose sooner or later. And when it did, the amount of strain released was unprecedented.

For half a minute that must have felt like a lifetime, 320,000 square miles (825,000 km²) of Central and North America shuddered and rumbled up and down and from side to side. More than twenty million people, some as far away as Los Angeles, Guatemala City, and Corpus Christi, Texas, felt the earthquake. Even though the rupture zone was thirty miles (50 km) offshore, south and west of Mexico City, the quake might as well have been directly underneath the downtown core.

Like the lowest notes of an upright bass, this fractured slab of sea floor played fatal music, a throbbing rhythm that pulsed with stunning efficiency through 190 miles (300 km) of continental crust to reach the capital city. The first burst of notes lasted roughly thirty seconds—about the duration of most normal earthquakes that occur closer to the surface. But then another segment of the plate broke loose and the vibrations started again. Many towers that had survived the initial attack were too crippled to endure the second.

As if that weren't enough, a significant part of the center of the city was constructed upon a layer cake thirty feet (9 m) thick of sand and gravel and clay, washed into the Mexico Valley from a ring of volcanic mountains. The modern city occupies the same site as Tenochtitlan, capital of the Aztec empire, built on the shore of an ancient lake called Texcoco. After Hernán Cortés conquered the Aztecs in 1521, Spanish engineers drained the lake to make room for a larger settlement. They knew nothing at all about the risk of massive earthquakes from the sea.

Four centuries later, eighteen million Mexicans had stacked a sprawling metropolis around and on top of this mudflat basin. When deep bass notes from the Michoacán gap hit the loosely packed soil, it throbbed and resonated. Apartment blocks on the dried-up lake moaned and swayed to the rhythm of the quake. Swinging side to side like inverted pendulums, each tower rang at its own frequency, depending on how tall it was. High-rise towers on the sands of Texcoco were in big trouble.

Every two seconds another shockwave thundered in from the coast like a horizontal pile driver hitting the foundations of the ancient city. Concrete and steel joints were cracking, walls were pulling away from floors. Tall structures standing too close together smashed into each other at the top and tore themselves apart from the roof down. Multi-wing complexes that were not perfectly square or rectangular—those with T-shapes or with a number of sections connected to a central hub—yanked themselves apart at the joints because each wing rattled and snapped to its own beat in syncopated self-destruction.

Anything between six and seventeen stories tall had a special problem—an internal rhythm in near-perfect synch with the earthquake itself. Shockwaves from the quake caused these midsize towers to hum like a tuning fork. This "resonant oscillation" amplified the amount of energy that pulsed through their vertical frames. With the arrival of each new wave, the foundations would be slammed to the side yet again before the rooftops had flexed back to a vertical position. Like rocks along the fault itself, the urban bricks, mortar, and steel inevitably failed.

As stress was relieved along the fault, the continental shelf off the west coast came unstuck from the oceanic plate that had snagged it more than ten miles (16 km) below the surface. The overlying crustal plate broke free, rebounded sideways as far as eight feet (2.5 m) and then sagged. Western beaches slumped as much as thirty inches (80 cm), causing the local sea level to rise.

Offshore, while the continental shelf was rattling loose and rebounding to the west, it also heaved upward. Roughly three thousand square miles (7,500 km²) of the shallow sea floor lifted a huge mound of salt water. When gravity broke this hump of hoisted water into elliptical waves—a train of tsunamis raced across the Pacific.

Given the magnitude of the earthquake, the size of the tsunami was relatively small. From Manzanillo to Acapulco and Zihuatanejo, the waves measured from three to ten feet (1–3 m) but caused relatively little damage. When the tsunami hit Hilo, Hawaii, hours later the wave was less than nine inches (22 cm) high. One possible explanation is that the ocean floor had been shoved underneath the continental landmass at such a shallow angle that the volume of seawater lifted by the earthquake did not amount to much, compared to other tsunamis generated by large earthquakes.

But in September 1985 scientists were still struggling to understand what had happened. The intricate details of tectonic motion were still pretty sketchy. One thing, however, was clear: a seemingly logical explanation for the lack of major earthquakes off Mexico's west coast—if we haven't seen any in all of recorded history, they must not happen here—was wrong.

Since a great earthquake had just happened in a supposedly aseismic zone like Mexico's, where else might the same thing occur? What about the coast of northern California, Oregon, Washington, and

British Columbia, also thought to be aseismic? A major disaster like Mexico's had not happened in the Pacific Northwest in all of recorded history either. By the same logic, if mega-quakes were likely to happen up there, surely we would have seen one by now.

On the other hand, maybe not. Overnight it seemed anything was possible. Perhaps a megathrust quake could happen on *any* offshore "subduction zone." Teams of scientists scrambled to Mexico as quickly as they could to examine this newest twist in the young science of plate tectonics. Ideas about how the world's largest earthquakes are created would change in the months and years ahead. For some scientists, the Mexico City disaster would be a tipping point.

As I fell asleep that night in New York, I too faced a knowledge gap. I did not know the Mexico City earthquake had become the opening chapter of a story I would soon be covering myself—a plot line I would follow for the next twenty-five years. It's a mystery that continues to unfold like a dimestore thriller, one that probably will not end for me until my home on British Columbia's coast has faced its own seismic nightmare.

CHAPTER 2
Lessons from the Rubble: A Front-Page Story

From the parking lot my first impression of the Pacific Geoscience Centre on southern Vancouver Island was that this sprawling complex of government laboratories belonged on the cover of some architectural magazine. The atrium was covered by a steep, angular skylight of tubular white steel and a grid of glass panels. Against a fetching backdrop of West Coast cedars and ferns, the mature landscaping included a small but brilliant reflecting pool. Nice digs for such a muddy-boot kind of science.

The downside to this idyllic locale was that it nestled cheek by jowl against the main runway at Victoria's international airport. Walking in to PGC (as it's known locally), I glanced at David Kaufman, the Toronto-based CBC producer working with me on a new documentary assignment to find out what scientists had learned from the Mexico City disaster. We both winced at the thought of random bursts of ruinous noise from turboprops, passenger jets, and helicopters that would no doubt make on-camera interviews here a challenge.

We were met at the door by seismologist Garry Rogers, who gave us the fifty-cent tour along a convoluted route past a museum-like display of earth science exhibits, presumably designed for public and school tours, to his office deep inside the complex. Once past the main public areas, the work space began to look more familiar, like a thousand other labs and universities.

Arriving at the top of a vast open stairwell, on a landing behind a plate glass wall, I saw a rack-mounted array of fifteen seismographs, each paper drum a furrowed field of squiggly parallel lines, pen strokes that looked like several days' worth of data—every little twitch of the ground under this mountainous corner of the world. Pinned to corkboards, stapled onto beige walls, and taped to nearly every office door were charts and diagrams, cutaway illustrations of subduction zones and volcanoes, aerial photographs of the coastline, long strips of seismograph paper smeared with erratic ink blots, and reprints of science posters.

We found an empty square table in the coffee nook that looked like the only space big enough to fit several scientists in a group interview. Three of the men we were about to meet were out-of-towners with no office space in the building, visiting members of a team of seismologists and engineers who had just returned from a fact-finding mission to Mexico City. Over the coming days and weeks they would jointly write a science paper for the Canadian government's committee on earthquake engineering, the panel of experts who make changes to the national building code.

"The most surprising thing for me," said Dieter Weichert, head of seismology at PGC, "was this two-second period of earth motion, which for an earthquake of this size should have been longer." He described the tectonic origin of those strong pulses of energy that had pounded through the earth with clockwork regularity every two seconds. Think of a metronome clicking back and forth or a relentless drumbeat.

It seemed to Weichert that a broken plate of oceanic crust ninety miles (145 km) long should have vibrated more slowly, perhaps once every six seconds. One might think the larger the slab of rock the slower the drumbeat, yet for some reason that's not what happened. The shockwaves in Mexico had come bang, bang, bang—every two seconds.

More fascinating to me, since I'd not heard any of this before, was the discovery that the deep soil of the dried-up lakebed underneath the city's downtown core had also vibrated every two seconds, which amplified the force of the incoming seismic waves. Who knew that soil had its own frequency and would quiver like a bowl of jelly? I tried to imagine evenly spaced ripples rolling through dirt like waves on the ocean.

It turns out the speed of vibration is determined by the type of soil (what it's made of) and how hard or loosely packed it is. Weichert explained that structures of a certain height (six to seventeen stories tall) can also have a natural vibration period of roughly two seconds—which amplify the shockwaves yet again. The image of a seventeen-story tuning fork came to mind.

The seemingly random damage pattern Robb Douglas had told me about on the phone from Mexico City back in September began to make sense. Before leaving on this shoot I had studied his news footage from the disaster zone: one condo block in ruins while another right beside it, maybe only a few stories taller or shorter, was left standing.

To my surprise, the Mexico City building code already included specific and more stringent regulations for anything constructed on the lakebed because there had been damaging quakes there in the past and local officials thought they knew what to expect. But Ron DeVall, a civil engineer in his midthirties who sat across from Weichert, pointed out that this bizarre, unlucky, two-second rhythm had never been seen before.

"They knew that the effect of the soil was pronounced in these areas and their code did account for it," confirmed DeVall. "On firm ground they were talking about basic acceleration of 4 to 5 percent of gravity." In other words, to survive the kinds of shaking that had hit Mexico City in 1957 and 1979, any new project constructed on bedrock since that time was in theory designed to withstand a lateral jolt against its foundations equal to 4 or 5 percent of gravity—meaning 4 or 5 percent of the building's own weight. That much lateral reinforcement was a fundamental requirement of the local code.

Anything constructed on the lakebed zone of the city was supposed to be even stronger, able to withstand twice as much lateral force—up to 10 percent of gravity. The shockwaves of September 19, unfortunately, had been much more powerful than that. "In actual fact," said DeVall, "some of the instruments were measuring 20 percent gravity. So they were getting a very large shake, much higher than they had anticipated."

DeVall compared the crumbled towers to a child on a swing. "If you can time the rhythm of your pushing, you can just drive that swing higher and higher and higher. In essence, that's what happens to a building sitting on the lakebed. The soil and the base rock were almost in resonance and the vibration was amplified to the building. The forces generated were much higher than what the building codes predicted. And the effect on the buildings was dramatic. It produced a much wider range of major damage than the previous earthquakes did."

Two other factors added to the toll of damage and destruction as well. Most of the Mexico City seismic regulations were written and passed into law as a result of the 1957 event and did not apply retro-actively to thousands of city blocks constructed before that. Tragically, this included many schools and hospitals built decades earlier. "Seeing a lot of hospitals and schools damaged, which are traditionally thought to be areas of refuge in a disaster," said DeVall, "to find them gone is, to my mind, a serious problem."

If there was any encouraging news from Mexico, it was that the damage was not evenly spread across the entire urban area, according to Robert Lo, a civil engineer who specializes in soil conditions. Lo even joked that on the day they had arrived in Mexico, roughly three weeks after the temblor, they hadn't seen any significant damage. "We thought we were in the wrong city," he smiled.

It soon became obvious that the majority of the urban area had been built on solid ground and that well-engineered buildings were able to ride out the shockwaves with relatively little damage. Ironically, a lot of older buildings not expected to survive—built of masonry and other materials considered quite brittle and susceptible to shockwaves—were still standing because they weren't tall and hadn't resonated with the vibrations coming from this particular quake. The airport remained open throughout the disaster, the metro subway kept running, nearby dams did not collapse, and most of the main waterlines survived the shaking as well. Even the electricity remained on in many neighborhoods.

But those ruined hospitals and schools had convinced DeVall that we in North America ought to learn something from Mexico's misery. "It might be prudent on our part to review these buildings to make sure that they will sustain an earthquake," he said. "You definitely want that hospital not only to be standing up afterwards, but functioning afterwards." He pointed out that New Zealand, another seismically active area, had already begun this kind of survey of essential infrastructure. The bad news was that, as far as he knew, no seismic inspection was planned for urban regions of the Pacific Northwest.

News reports at the time said the Mexico City rupture had been unexpected. What I wanted to know was why. Aren't all earthquakes unexpected? "Well," explained Dieter Weichert, the Mexico disaster "was a bit of a surprise because they had a historic record of two hundred years without a large earthquake. And there was reason for thinking there might not be one." True, several smaller shocks had damaged Mexico City in the past and yes, there had been ruptures to the north and south of the plate that slipped this time. Through all the previous rumbling and lurching, however, this one segment of the ocean floor had remained quiet and nobody knew why. Could one subsection of the ocean floor be so different from the broken fragments on either side? To me this sounded like the missing clue at the heart of a great mystery.

The unexplained absence of large quakes over a long time seemed to be the soft underbelly of the "aseismic subduction" hypothesis—the conventional wisdom that the jagged edges between some tectonic plates are "not like the others" and don't get stuck together. Weichert told me that some parts of the ocean floor were younger and hotter rock than the overlying continental crust and perhaps that was why this slab had been aseismic. The interesting twist was that in the aftermath of Mexico City, at least for the scientists gathered in this room, there was reason to believe something might be wrong with conventional wisdom—at least the aseismic part of it.

"People thought that the plate was subducting smoothly," said Weichert. And yet, quite obviously, it had been stuck and building up strain for two centuries. Seismologist John Adams from the Geological Survey of Canada's earth sciences lab in Ottawa told us that a tectonic plate very much like the one that wrecked Mexico City—only three times larger—lies just offshore west of Vancouver Island, extending down the outer coastline of Washington, Oregon, and northern California.

"It's been demonstrated that the Juan de Fuca plate is subducting under North America. But we see no earthquakes along it. And we have a historic record of 150 or 200 years without large earthquakes. Therefore, there are two possibilities: either it is smoothly sliding under North America, doing it continually and without large earthquakes, or strain is building up for a large earthquake—of a type that would only happen every 200 to 500 years."

Was there an echo in this story? I was hearing for the first time about a slab of Pacific Ocean floor called the Juan de Fuca plate, which runs from Vancouver Island to northern California and is sliding eastward underneath the continental plate of North America. This eight-hundred-mile (1,300 km) fault, the boundary line between two tectonic plates, will later be named the Cascadia Subduction Zone. As I listened to John Adams, it sounded exactly like Dieter Weichert's description of the plate that got stuck sliding underneath Mexico. Until September 19, both fault systems had been thought of as aseismic.

I didn't know it at the time, but John Adams had already been studying this mystery for five years. He had published three papers on faults that lie deceptively silent for hundreds of years, including an alpine fracture zone in his home country of New Zealand. Now, in one long breath, he had just warned us that what we saw in Mexico City might also happen in the Pacific Northwest. And Adams wasn't the only one to say it.

Garry Rogers was the young PGC seismologist whose job it was to monitor the two hundred or so temblors that rattle through southwestern British Columbia every year (only two or three of which are strong enough to be felt). He told us the historical lack of huge megathrust events off the west coast of Vancouver Island could be very misleading.

"The implication," said Rogers with focused intensity, "is that the possibility for very large earthquakes—the kind that occurred in Mexico just recently—does exist on the west coast of Canada. The problem is that in the 150 short years that we've been here, we have not seen any examples of earthquakes on our subduction zone. Not even small ones."

He explained that those two hundred rumbles occur because of stress within the overlying crustal plate, relatively close to the surface. The much larger shock—if it does happen—would occur almost forty miles (65 km) below ground along the length of the subduction zone. Like Adams, Garry Rogers thought the absence of deep Juan de Fuca quakes put seismologists in a quandary.

"At the moment, we just don't know," he said. "It's a subject of scientific debate. But if we compare other areas around the world that are very similar to our subduction zone, we find that we are the only one that has not had large earthquakes."

For seismologists in 1985 it was hard to imagine why the Juan de Fuca plate (or the Cocos plate in Mexico) would be special—the only place on the planet where two plates glide past each other trouble free. How could this not be like the dangerous and deadly subduction faults off the coasts of Alaska, Chile, and Japan? Although Rogers didn't seem like a gambler, he was willing to speculate.

"A more likely scenario, comparing it with other zones, is that we *are* capable of large earthquakes but with very long intervals in between them," he said. The long quiet history of Juan de Fuca could mean "it's stuck and one of these days we're gonna have a monster earthquake like Mexico had."

If the fault were "stuck," I wondered, could the build-up be measured and—if you could *see* the stress increasing—would it be possible to predict the next quake? "It may be," answered Rogers. "And, in fact, one rather suspects it should be, because before such a large earthquake a tremendous amount of strain is stored up. We might be able to detect a deformation like that. In fact, they can see this kind of thing in Japan since their last big earthquake—deformation going on."

Evidently rocks bend and tilt under stress and there are changes in electrical signals coming from the earth, all of which could be monitored. Rogers described prediction as a dark art that was still many years away from success, but his point was that there are things we could and should be doing to confirm or deny the possibility of large subduction earthquakes off the Pacific Northwest coast.

It turned out that John Adams was already doing exactly that kind of research. Less than a year earlier, while working at Cornell University in New York State, he had published the first in a series of papers with new data showing that the coastal mountains of Washington and Oregon were in fact being bent and tilted landward, probably by the force of plate tectonics.

A magnitude 8 or higher, *here* on *my* West Coast—really? I'd been living in Vancouver nearly ten years at that point and had never heard anything about a monster shockwave. Not a word of it. How could I, a working journalist covering British Columbia for most of a decade, have missed a blockbuster story like that? Well, it turns out the banner headline was being written in the present tense at that very moment. This news had not escaped the confines of laboratory walls until now.

With a quickening pulse, I turned back to Dieter Weichert and asked for context. He recited what sounded like a well-rehearsed list of the most recent moderate-size temblors in the Pacific Northwest: "For ten years, we've always warned people that there are earthquakes—in Seattle-Tacoma [in 1965], under Pender Island in 1976, central Vancouver Island in 1946." It was true, he conceded, "We have never talked about this big subduction earthquake. We knew about the possibility, but certainly with a fifty–fifty chance, you're not going to say there is a big earthquake waiting for us."

First, I asked myself, who else knew there was even a fifty–fifty chance of a magnitude 8 rupture? Probably nobody except the scientists. Then it occurred to me—okay, so the senior seismologist at the Canadian government's West Coast geoscience laboratory is a cautious man who doesn't want to alarm the public without reasonable and probable cause. I understood that. Yet now, in the aftermath of Mexico City, he was apparently ready to raise the biggest, reddest warning flag I'd ever seen.

"Now you're saying it?" I prompted.

Weichert took the plunge: "We're saying yes, we have to come to grips with this problem. The chance has increased, in our minds, from a fifty–fifty chance to something like a seventy–thirty chance *for* the earthquake to happen within, say, the next two hundred years."

As a scientist, he really couldn't say for sure when the megathrust might happen—two hundred years from now, or *tonight*—so Weichert had erred on the side of caution. That's what responsible government scientists do. Kaufman and I, however, figured Weichert, Rogers, and Adams had given us a clear signal that the risk level was sufficiently high to justify front-page treatment of the issue.

On Sunday, November 3, 1985, I flew from Vancouver to San Francisco en route to the U.S. Geological Survey laboratory at Menlo Park, California. First thing Monday morning we shot an interview with

USGS seismologist William Bakun, who not only reinforced what the Canadian team had told us the previous week but made an even more ominous prediction. He said the Juan de Fuca plate could generate a disaster even larger than the one in Mexico.

"We have to take seriously the possibility that a great earthquake—a very great earthquake, such as the 1960 Chilean earthquake—might occur along the Washington, Oregon, and British Columbia coast," said Bakun. "We're talking about as big an earthquake as has occurred in historic time—in the world."

Knowing almost nothing about what happened in Chile, twenty-five years earlier, I again asked for clarification. "Where would that be on the Richter scale?"

"Off it," he laughed weakly, and then quickly followed with an explanation. A moderate earthquake is defined as magnitude 5.0 to 5.9; strong is 6.0 to 6.9; major is 7.0 to 7.9; and a *great* earthquake registers 8.0 or higher on the Richter scale.

Because the scale is logarithmic, there is a tenfold increase in the amplitude of the shockwaves with each higher whole number on the scale. If a magnitude 4 caused rocks to vibrate and move less than half an inch (1 cm), a magnitude 5 would cause them to move four inches (10 cm). Some studies have estimated that this tenfold increase in the amplitude of the shockwaves would require thirty-two times more energy. So a magnitude 9 would generate thirty-two times more energy than a magnitude 8.

The Mexico City quake was an 8.1 and the 1960 Chilean disaster was a 9.5, the largest temblor ever recorded by modern instruments. That means the Chilean rupture generated more than thirty-two times the energy of the Mexico City event. And here was William Bakun of the USGS telling us to expect the same in the Pacific Northwest.

We had come to Menlo Park primarily because Bakun and his colleague Allan Lindh had recently launched the first high-profile earthquake *prediction* experiment on U.S. soil. The Chinese and Japanese had both been running prediction studies for several years already, but given their spotty results and the controversial nature of spending money to forecast disaster, this was a bold leap for the USGS. As a journalist I figured the first thing people living in any hazard zone would want to know was: *when* will the Big One finally happen? Now some of America's top scientists were trying to provide an answer.

"We can predict earthquakes, in one sense," Bakun said, cautiously. "We can identify sections of plate boundaries that will eventually fail in large, damaging earthquakes." Figuring out *where* the San Andreas fault might break again, or being pretty sure that the Juan de Fuca plate will rip loose from the North America plate *some day*, sounded like important science to me, although I'm pretty sure that's not what most people think of as prediction. Bakun agreed. "We still do not know how to predict earthquakes on a short-term basis. That has turned out to be a very difficult problem, and it's a focus of our ongoing research."

Bakun's coauthor in the prediction study, seismologist Allan Lindh, told us they were studying a stretch of the San Andreas near the farming town of Parkfield, California, that had ruptured five times since 1857—each event a magnitude 6 temblor that seemed nearly identical to the one that came before, as if the same punch were being thrown over and over again. Bakun and Lindh had convinced themselves the next in this series of "characteristic earthquakes" was due in about three years. According to their calculations, the fault would build up enough stress to break again as early as January 1988.

In August 1985, only a few months before our visit, Bakun and Lindh published the first official seismic prediction ever issued by the USGS: "The next characteristic Parkfield earthquake should occur before 1993." Even with a five-year fudge factor, they had stuck their necks out by putting the prediction in writing in *Science*, one of the most prestigious and high-profile research publications in the world.

Like Bakun, Lindh seemed to be a cautious man. Still, there was enthusiasm in his voice as he talked about trying to trim the fudge factor and "narrow down the time from a few years to months, to a few days." He told us, "I think we've got a fighting chance," asserting that the way to refine the prediction was to concentrate as many instruments as possible along one small segment of the fault—the same fifteen-mile (25 km) rupture zone that had moved in each of the previous Parkfield punches—and monitor every little creep and twitch in the earth, day and night, until the next rupture. With luck, they might spot some kind of precursor that would allow them to issue a warning to the public.

When we wrapped the USGS shoot late that afternoon, my team and I drove four hours south on Highway 101 from Menlo Park, through the rush hour of San José, to a wine-country town called Paso Robles, where we spent the night. Early next morning we headed east into ranch country across dry brown hills on bumpy two-lane blacktop in search of a wide spot in the road called Parkfield. Our map showed it smack in the middle of nowhere about halfway between San Francisco and Los Angeles.

Roughly thirty miles (50 km) east from Paso Robles we passed a road sign that told us we had found what we were looking for. Parkfield had a population of thirty-four, not counting cattle, and stood 1,520 feet (463 m) above sea level. We pulled up in front of what looked like an old ranch house made of square-cut timbers the color of creosote with a wide veranda, a corrugated metal roof, and a big stone chimney. Out front, in a tidy patch of unnaturally green grass, stood a tall wooden cowboy carved from a log and bolted to a stump with a small wooden dog at his knee.

In a gravel parking lot stood the rusty iron hulk of what used to be a water tower. In a curvy flourish of creamy white letters, a hand-painted sign read "The Parkfield Cafe." Under that, in slightly smaller print, was the proclamation "Earthquake Capital of the World. Be Here When It Happens."

Farther down the road we found the man we were looking for. USGS technician Rich Lichtie was waiting for us beneath a one-lane bridge that spanned a gully where Little Cholame Creek trickled west toward the sea. Lichtie fit the landscape in his baseball cap, blue chambray work shirt, jeans, and cowboy boots. His windburned complexion and red walrus moustache allowed him to blend in with the surroundings even better. No labcoated scientist from the big city, Lichtie was the guy in charge of all the USGS equipment that was jammed into the ground along both sides of the fault, and he clearly spent a lot of his time outdoors.

Of all the places he might have arranged to meet us, he had chosen this bridge for a reason and wanted us to see it from below. So we unpacked our gear and trudged down into the gully for a closer look. That's when Lichtie explained that this big ditch was part of the San Andreas, that the bridge literally crossed the fault, and that the last Parkfield event, back in 1966, had torn the old bridge off its foundations.

We were looking at a replacement span that already showed signs of stress. Doug, my cameraman, got a telling close-up of one big bolt holding two heavy steel girders together by no more than a few threads. The two main sections of bridge deck had already been pried apart far enough for sunshine to burn through a gap between the beams. It was a crude yet graphic display of creep along the fault.

Lichtie took us up the road to a cow pasture, where we hiked across the dun-colored grass toward a dry gulch with a storm culvert dug vertically into the earth. When he removed the cover, we could see a metal platform bolted to the corrugated wall of the culvert with a cable-and-drum contraption that looked like something a kid might build with an erector set.

Halfway down the culvert was a circular hole cut into the earth several feet below the surface, where a horizontal plastic drainpipe extended toward the far side of the gulch. Inside the pipe was a pencil-thick braided steel cable that looked like a buried trip wire. Lichtie called it a creepmeter and explained that it was pretty much what it looked like—a wire sixty-five feet (20 m) long, stretched across the fault and connected to a strain gauge (in a toolbox at the bottom of the culvert) capable of measuring even a few fractions of an inch of slip along the plates.

Next Lichtie took us to a nondescript shed in a grove of walnut trees, halfway up the side of the gulch. When he unlocked the door we saw what looked like a high-end amateur telescope, with a white steel barrel the size of a small cannon, mounted on a high-tech tripod anchored to a concrete pad with a small spotting scope bolted on top like a rifle sight. He switched on the power and a cherry-red laser shot a beam across the valley toward another tiny shack so far in the distance we could see only a smudge through waves of dusty heat rising in the noonday sun.

Lichtie used the rifle scope to line up the laser with a parabolic reflector in the other little shed three miles (5 km) away. "We're shooting the beam across the fault to a reflector, which brings it back here. And we can measure to within a half a millimeter how far that reflector has changed in relation to this building," said Lichtie. As expected, the laser device had already documented right-lateral motion along the fault—the Pacific plate creeping north toward Alaska.

As part of Bakun and Lindh's experiment, the USGS was in the process of installing a cluster of these and other instruments at various points along the fifteen-mile (25 km) rupture zone in Parkfield. The data were being beamed continuously by microwave to a real-time processor in Menlo Park, where members of the research team were keeping constant watch. They even wore pagers that would wake them in the dead of night or ruin a perfectly good dinner if the fault started to creep or warp or bend itself out of shape.

When producer David Kaufman and I realized the fault ran right up the middle of this gulch, it was impossible to resist the temptation to straddle the fracture and take a picture. Not that you could really see a crack or crevice in the ground; there was nothing more to look at than the V-shaped bottom of this little gully, swathed in straw-colored pasture grass in dire need of rain.

A better way to see the fault was from the air. Aerial pictures showed that one side of the fault had been thrust up slightly higher than the other side, enough to cast a distinct line of dark shadows that ran for miles and miles in the early morning light. There it was, plain as day—two tectonic plates grinding past each other at the blistering speed of two inches (5 cm) per year, roughly the same speed as your fingernails grow.

Far above the valley floor, it was also easier to see why the prediction experiment was being conducted along this particular segment of the fault. The San Andreas is not a straight line. Far from it, especially in Parkfield, where the shadow line zigzags ever so slightly and in a stretch northwest of town actually kinks. There's a five-degree bend along a segment 1.2 miles (2 km) long that was the epicenter of the 1966 quake.

In their paper, Bakun and Lindh referred to this as a "geometric discontinuity" and suggested the bend probably controls how much of the fault moves when it ruptures. I imagined it as a kind of plug or doorstop jammed in the crack, causing the fault to slow down or temporarily stop moving. Not only that, with all this zigzagging, there are rough patches in the rocks—seismologists refer to these as asperities—that create friction and also slow the creeping motion along the fault. While the rest of the San Andreas is ripping along at almost 2 inches (5 cm) per year, the Parkfield segment is lagging behind at only 1.4 inches (3.5 cm) per year.

With tectonic plates as big as these, however, it's obvious the little snags—those Parkfield asperities—can't hold up progress forever. The stress builds to a point where the rocks fail. The rough spots finally shear away. In the span of less than a minute, roughly twenty-two years of "lost motion" along the fault is recovered as the Parkfield segment catches up with the rest of the San Andreas in a shuddering leap to the north. An earthquake.

Another important reason for clustering so many instruments here is that just before the 1966 main shock, the earth may have given off subtle warning signs. Twelve days before the temblor, fresh cracks appeared in the ground near the center of the rupture zone. Nine hours before the main shock, an irrigation pipe that crossed the fault broke and separated. Were these true precursors? Maybe. That was still the subject of vigorous scientific debate. But even the outside chance of a successful prediction was enough to motivate the USGS team to do everything they could to spot reliable symptoms of the next San Andreas quake.

CHAPTER 3

The Alaska Megathrust: Cascadia's Northern Cousin

Even before the earth hammered Mexico City there were telltale signs of what to expect from Cascadia's fault. One of the first clues arrived by sea at ten minutes past midnight on Good Friday, March 27, 1964. A tumbling ball of water moving southbound from Alaska at an estimated 330 miles (530 km) per hour passed beneath the hulls of ships at sea without causing a stir.

At surface level in the open ocean, it felt like just another swell in the North Pacific. But this was a "seismic sea wave," what used to be called a tidal wave (now known as a tsunami) and it was very different from the chop and rollers left by a storm that had passed through a few days earlier. To a sailor's wary eye, only the three-foot (1 m) crown of this monster's head would have been visible that night, just another hump in a sea of thousands, with all its furious strength hiding in darkness below.

Unlike ripples, whitecaps, and windblown breakers that churn only the surface, this rolling mountain of brine reached all the way to the ocean floor and traveled at the speed of an airliner. When it reached the western side of Vancouver Island, the front of the wave began to slow as it scraped over the shallow bottom of the continental shelf, forcing the back half to mound up eight feet (2.4 m) above a normal high tide. An edge of the wave then sheared away and made a fast left turn into a fjord called the Alberni Inlet.

As the turning flood crashed over rocks at the Bamfield lighthouse on the outer coast, a keeper on duty grabbed the phone and placed an urgent call to Port Alberni, a mill town at the head of the inlet. The funnel shape of the inlet itself—a saltwater canyon cut through narrowing mountain walls—squeezed and amplified the wave as it shot toward the heart of Vancouver Island like a cannon ball. The fast-rising swell ran forty miles (65 km) from the Bamfield light to the head of the fjord in only ten minutes, not nearly enough time to spread the word. Port Alberni might as well have had a bull's-eye painted across its industrial docks, marinas, and low-lying residential streets.

No one noticed the fist of frigid seawater as it lifted two channel marker buoys and thundered across the threshold of the inner harbor. Night shift longshoremen, completely unaware, continued to hoist and sling bundles of lumber aboard the *Meishusan Maru*, a Japanese freighter at the sawmill dock. In the nearby pulp mill, boilers were running full steam. Paper machines were spinning out massive rolls of newsprint for the *Los Angeles Times*.

With little to do after midnight, most people in this town of seventeen thousand had already gone to bed. Only a handful heard skimpy stories on the late night news about an earthquake that had rattled Anchorage 1,120 miles (1,800 km) away. Even those who did hear about the jolt up north would never guess what was about to happen in the Alberni Valley or how it was connected to Alaska.

Crossing the harbor at 240 miles (386 km) per hour, the tsunami surged beneath acres of floating logs, breaking boom chains, snapping steel cables, and scattering dozens of rafts of heavy timber across the inlet. It assumed the shape of a blitzing storm tide rather than a towering curl. As in Sumatra and Thailand many years later, the wave that slammed Port Alberni looked more like a river run mad than the perfect breaker that surfers catch only in their wildest dreams.

Minutes later the main water pipeline to the pulp mill broke like a twig. The *Meishusan Maru* rode the surge to the end of her mooring lines, twisted free from the sawmill dock and drifted toward a nearby mud flat. Bundles of lumber floated off the dock and rode the churning froth into downtown streets like so many cubic battering rams. When the leading edge of the drenching brine finally reached the head of the inlet, the last of its energy was spent running upstream against the Somass River. With catlike stealth the water slipped over the low dike along River Road and spilled into the bottom-land housing on the other side.

All these years later it's hard for survivors to recall what they heard first—the mysterious gurgling beneath their floorboards, or the heart-stopping thump of a mill worker's fist against their doors in the dead of night. Allan and Jill Webb lived in one of those houses near River Road. Forty-five years after that Good Friday night, they gathered with other Alberni tsunami veterans in the mayor's office at city hall to recall their experiences. Jill told me it was the awful hammering that she would never forget.

"Well, that's what I think woke us up," said Jill. "You know, the banging on the door." But she never saw the man who spread the alarm because as she climbed out from under the covers, her feet plunged into frigid wet muck and her attention went straight to the floor. "I stood up in this water and I thought 'Oh, what is this?'" She laughed. "It was quite a shock to step into this water."

Her husband, Allan, recalled the sequence of events slightly differently. For him the sound of trickling came just before the knock. "I got out of bed and stepped into about six inches of water. So I'm thinking, 'Well, is there a pipe leaking?' And of course, we have no idea what this is." But then he glanced out the window and in clear moonlight saw what appeared to be a lake rising all around them. Then he spotted the lone, unidentified Good Samaritan.

"He was running down the alley," recalled Allan. "And by the time he got to the end, the water was above his knees, so he couldn't hang around too long." Then they heard a great roar as boilers at the pulp mill blew off steam. Workers were desperately trying to bleed away pressure before cold seawater hit hot pipes and caused an explosion. The mess inside the Webb's house was by now three feet deep and rising as Allan and Jill made their way to a small side room where their twenty-three-month-old daughter, Carrie, slept.

They found Carrie standing in her crib, staring wide eyed but silently at the gush coming up through a trap door in the floor. Allan estimated the flood had risen another foot within minutes. "I got a drawer out," he said with a shrug. "I figured I'd float the child around. And Jill—she can't swim, so I don't know what I'd have to do with her. But . . ." He let the story fade.

Jill was still remembering the look on little Carrie's face. "She was just sort of standing up looking—you know? With big eyes. And we just picked her up," said Jill, "but she was really very calm—calmer than we were."

Yvonne Forbes, a neighbor down the street, recalled that her father, who worked at the mill, grabbed a ringing phone that woke everyone in the house. "Somebody called him from the mill and said there was a tidal wave, but he didn't really believe it," she said. "I mean, from an earthquake in Alaska?" She mugged a look of disbelief. "It just sounded like a fairy tale, really. You know? And nobody knew anything, really." Heads around the room nodded in agreement. Then she continued her father's story. "It was in the middle of the night, pitch black, no power. So he opened the back door—I don't know why—and then he could hardly get it shut. And the water just was coming in."

When Yvette Gaetz started telling her story, every mouth in the room fell open. "Well, I was nine months pregnant," she said, matter-of-factly. "My baby was due that night." She and her husband, Simon, also had three other children under the age of three asleep in the back bedroom.

"My uncle worked at the plywood [mill] and he lived next door. And when the plywood sent him home, he just phoned and he said, 'Yvette, there's a tidal wave!' And I could see water coming under the kitchen door. So, as I was talking to him, I say, 'I gotta go, uncle. The water is comin' in!'"

When Simon spotted the dark, muddy scum spreading across the kitchen floor, he rushed to the front of the house and flung open the door, only to face a larger, blacker torrent. He and Yvette's brother Ray, visiting from Saskatchewan, threw their combined weight against the incoming flood.

"We had a hell of a time closing the door," said Simon. "We got it closed and that's when we started putting everybody up in the attic." It seemed like the only way out was going to be up, so they grabbed a chrome highchair and planted it under the trap door to the attic. Yvette lifted the youngest child from a crib while Simon and Ray gathered up the other two children.

"I'm left in there with the baby, and I gotta get my baby out. Now, I don't know how I did it, but I crawled on top of that highchair—being nine months pregnant and holding a thirteen-month-old in my arms," she paused for a breath, rolled her eyes and went on. "But then I see the fridge. It was *floating* in front of me—with the plug still in the wall!" There was a sudden intake of breath as everyone in the room got the picture. "And I thought, 'Now where do I go?'" Yvette laughed and the room laughed with her, nervously.

Simon and Ray put the three-year-old child on the couch and almost immediately it lifted off the living room floor and began to drift. They put Marcel, the two-year-old, on a mattress and the mattress started floating as well. "Marcel was fine as long as he didn't move—but he was two!" said Yvette, her voice an octave higher.

Then, because there was no ladder in the house, Ray stood on a chair and crawled up into the attic first. With Ray pulling and Simon pushing, Yvette managed to climb up and squeeze through the small square hole leading to the attic. When she got herself turned around and stable, Simon started passing the other children up to her. By now she was drenched and freezing cold, so the last thing Simon did before climbing up himself was to grab a handful of clothing.

"So the kids were dry, but I was soaking wet," she explained. "And he just grabbed an armful of men's clothes, because that was the closest thing. And there I am changing clothes and hoping and praying the water wouldn't come up." She looked around the room. Her audience was spellbound.

"But my brother, I guess he was—he was more worried about me, thinking I was going to go into labor—because I was due that day! 'What are we going to do if she goes into labor here?'" Yvette shook her head at the absurdity of it all. "But everything—nothing happened. I mean, the Good Lord was with us, I guess."

As suddenly and mysteriously as it came, the flood turned around and started to go the other way. Jill Webb heard a funny sound and opened the living room window. She and Allan saw their car turned on its side, bobbing away in the moonlight. Carrie's baby buggy had drifted across the street with the receding current—on its way west toward the ocean.

"It was really a very, sort of eerie feeling," said Jill. "It sucked out." To underline the point, she created her own sound effect. "It just went *shooooop*—you know?—as it was leaving." She shook her head. "It was really kind of an awful sound . . . Fortunately it didn't take us, too."

But that was only the first of six waves to come that night as the tsunami oscillated back and forth down the inlet like soapy swells in a giant bathtub for eighteen hours. Wave number two, which arrived ninety-seven minutes later, was the biggest and most destructive. At 1:20 a.m. a ten-foot (3 m) surge pounded the city like a wrecking ball, picking up all the debris left by the first wave—fishing boats torn free from their docks, floating cars and trucks, buoyant bundles of lumber, thousands of busted-loose two-by-fours and thousands more raw logs, many weighing several tons apiece—and hurled these projectiles into the low-lying streets of Port Alberni.

As the turbulent seawater climbed the government tide gauge at the rate of one foot (30 cm) per minute, the crew aboard the *Meishusan Maru*, which had been grounded on a mud flat by the outgoing rush of the first wave, quickly fired up their engines on the second swell, got the freighter off the mud, turned it back toward the main navigation channel, and dropped anchors in a deeper part of the harbor. At the same moment, River Road houses began to float off their foundations. An eyewitness told a newspaper reporter the next day of seeing a large, two-story house drifting down the Somass River. It gradually broke up and sank. At an auto court near the riverbank, the rising swell lifted a row of six small cabins simultaneously.

Mary Rowland, another of the white-haired survivors sitting on a couch in the mayor's office, told of seeing her neighbor's house being swept away. "Joy Smith had a little store at Beaver Creek and River Road, and she lived across the street in a house. And she had a habit of always having a cup of coffee on the go. It sat on the end of the stove. And so their house went down River Road and into the fields—oh, maybe about three blocks or something—"

"The whole house?" I interrupted.

"Yeah, and the cup of coffee was still sittin' on the stove with the coffee in it." She patted her hands together with a tiny smile. "That's how calm it was. It just picked it up, took it along, and sat it down."

A disaster report from British Columbia's Provincial Emergency Program said it was hard to understand why no one got killed. The period of grace between the first surge and the disastrous second was not long enough to get everyone moving toward higher ground. "Many were caught in their homes," said the report. "The fast-rising waters knocked out all power and street lighting, so that many waded chest-deep, in the sudden dark, through their yards to safety . . . Even more miraculous were some of the hair-breadth escapes of children. One man dashed out to save his brand-new convertible only to find a pair of youngsters floating by on a log; he too was chest deep before the trio made it to dry ground. A civil defense worker rowing around in the dark checking houses, flashed his light into one and rescued a baby floating on a mattress."

Scientists would later confirm that this runaway train of six tsunami waves had rumbled through the night the entire length of the Pacific Ocean from north to south. Four children camping with their parents near a beach on the Oregon coast were swept out to sea and drowned. Northern California got hammered when four of the big swells dragged across the shallow ocean bottom, sheared off, and circled back from the south to smash the seaside town of Crescent City, killing another dozen people. Sadly, several of the deaths in Crescent City were caused by ignorance. In 2008 Bill Parker, a retired civil defense coordinator who had tried his best to get everyone to higher ground, explained what happened when the salty sludge started to recede.

"Everything was a mess," Parker recalled. "The streets were loaded with debris. Buildings were shattered; cars were on top of each other." Then, as the second wave began to ebb, people wrongly assumed the danger had passed. Curiosity and bravado drew them to the inky rubble.

"Five people lost their lives because they went back to get the money in their store," Parker explained. "The father said, 'Well, it's my birthday. Let's take a drink for my birthday.' And so they poured drinks and wished him a happy birthday." Just then the third wave struck, a twelve-foot (3.7 m) wall of water that killed the five who thought they'd been lucky. "If they'd left three minutes before," said Parker, with a slow shake of his head, "they'd have been safe."

Twenty-four hours later those six liquid mountains from the Alaska coast had finally splashed themselves apart against the icy shelf of Antarctica. At sunrise the next morning, through heavy mist and fog, Port Alberni looked like it had been ripped apart by tornadoes and then drowned. Fifty-eight houses had been washed off their foundations and destroyed, with 375 others severely damaged and a thick layer of silt and mud coating everything in sight. Chaotic piles of lumber and logs blocked roads and railway tracks. Upturned boats and cars were strewn everywhere, some cars piled atop others, one memorably parked on its nose, the front bumper buried in several feet of salty, sticky muck. Like the pedestal of some perverse art project, another car balanced a fifty-foot (15 m) log across its roof.

Two log booming boats were dumped high and dry on a downtown street and all of MacMillan Bloedel's industrial plants—the pulp and paper mill, the sawmill, the planer mill, and the plywood plant—were knocked out of service and shut down. The shape of Alberni's long, narrow canyon, cutting through steep mountain rock from the open ocean to mid-island, had focused the energy of the wave to devastating effect. Other island communities had also suffered damage—eighteen homes in the Hesquiaht village of Hot Springs Cove were wiped out—but no place on the British Columbia coast was as spectacularly messed up as Port Alberni.

By noon on Saturday, help was on the way as roughly two hundred soldiers and a detachment of the Royal Canadian Mounted Police approached the Alberni Valley. Dozens of refugees gathered in emergency shelters while civil defense workers and the Salvation Army served hot meals and coffee. Counting the losses from the largest tsunami to hit Canada's west coast in modern history, officials discovered that only a few people had been injured and none had died.

The residents of Port Alberni found themselves miraculously alive. But with no electricity and the local radio station knocked off the air, they had no news about how much worse things had been for their neighbors to the north. On the Alaska coast, where all this violence and destruction had begun eighteen hours earlier, it had been a night of death and destruction by land and by sea.

∞

The earth tore itself apart fourteen miles (22.5 km) underground in Prince William Sound, eighty miles (130 km) east-southeast of Anchorage, starting at 5:37 p.m. on Good Friday. A complex fault that no one could see broke from its epicenter in two directions at once—to the southwest and the southeast. The ground shook hard for at least four minutes as the rupture spread over a distance of five hundred miles (800 km).

Exactly how long the violent tremors lasted was hard to tell because "every seismic instrument within a radius of several hundred kilometers was thrown off scale" after the first few seconds, according to a U.S. federal study. How long it *seemed* to last depended upon who was telling the story and what kind of ground they were standing on at the time. If they were anywhere near the epicentral region, the lurching and jolting went on for at least four minutes and possibly as long as six. It must have felt like forever.

Measurements were made by a network of seismic stations around the globe, but again—because none of the instruments on the ground in Alaska had survived long enough to capture an accurate local record of the energy—there were discrepancies in Richter scale calculations as well. A report prepared by the U.S. National Research Council lamented that the seismographic record of this earthquake was "woefully incomplete." Some seismologists figured the magnitude was 8.3; others pegged it at 8.6. Part of the problem was the Richter scale itself.

As one geologist explained it to me, no seismograph in the world at that time could get a correct magnitude because—until digital seismographs came along—the entire earthquake spectrum could not be accurately recorded for any event that lasted longer than about a hundred seconds. Great earthquakes like this one can take as long as three hundred seconds, sometimes even more, to finish rupturing. At an average velocity of just over 1.8 miles (3 km) per second, that's how long it would take for a fracture like this to unzip over such a long distance.

Basically, the old Richter scale was good only for shocks up to about magnitude 8; anything bigger (or longer lasting) was off the scale, so to speak. So an extension of the scale—the moment magnitude scale—had to be devised for measuring the relative sizes of the 8+ events. Thus the Alaska temblor was eventually assigned a magnitude of 9.2 while assessment of the Chile quake of 1960 shifted from 8.9 to 9.5, the largest earthquake ever recorded with modern equipment. The point being that no matter how the numbers are crunched, the Good Friday earthquake was then and remains the largest scientifically documented seismic shock to hit continental North America and the second largest in the world.

Piecing together a sequence of events in the days and weeks after the Easter weekend, scientists concluded that the zone of significant damage covered 50,000 square miles (130,000 km²). The vibrations were felt over an area of 500,000 square miles (1.3 million km²). The pulses of energy had traveled roughly 1,200 miles (1,930 km) to the southeast, when a group of scientists attending the annual conference of the Geological Society of America, who were enjoying dinner in the revolving restaurant atop Seattle's Space Needle, felt the tower vibrate slightly.

Barometers in La Jolla, California, roughly two thousand miles (3,200 km) from Anchorage, detected an atmospheric pressure wave generated by the quake. Water levels in 650 wells across North America, in Hawaii, and as far away as South Africa jumped abruptly, one as much as seventeen feet (5 m). The Council report said "probably twice as much energy was released by the Alaska earthquake as by the one that rocked San Francisco in 1906." Measured by the newer moment magnitude scale, Alaska 1964 was 160 times larger than San Francisco 1906.

In the immediate aftermath of Good Friday, however, reports of casualties and property damage were slow in reaching the outside world because power and telephone lines were down all over south-central Alaska. Only those living through the disaster knew how bad things really were. The statewide death toll, at 115, would be described officially as "very small for an earthquake of this magnitude." Of those who died, 106 were killed by the tsunamis. But in some ways, the people of Alaska had shared a bit of the luck that blessed their neighbors down in Port Alberni.

Consider the element of timing. The rupture happened late on a holiday Friday afternoon, the beginning of Easter weekend, so schools were empty and most offices were deserted. Construction crews working on the new Four Seasons building in Anchorage had packed up and gone home only thirty minutes before the earth began to shudder. The unfinished and fortunately unoccupied high-rise collapsed in a heap of buckled concrete and twisted steel.

At the moment of the rupture hundreds of trollers and seiners were tied up and dozens of canneries were shut down because fishing season had not yet opened. As luck would have it, the tide that afternoon was among the lowest of the year, so the reach and hydraulic impact of the tsunami was somewhat reduced. Imagine what might have happened along the docks and boat harbors in the middle of a busy workday on a swollen tide in high season.

Although snow blanketed the mountain slopes and still covered the ground in many places even at sea level, the temperature had been pleasantly mild that afternoon, with highs near forty degrees Fahrenheit (about 4°C). But weather-wise Alaskans were still dressed for the most part in warm clothing, which no doubt helped them survive a long, cold night outdoors or until they made their way to rescue shelters. Of all the things that could have gone wrong, not all of them did.

Alaska was and still is a sparsely populated frontier; an event of this magnitude in southern California would have been catastrophic in terms of deaths and injuries. But to those who survived, numerical comparisons are meaningless. No matter how lucky they were, most Alaskans were devastated by the earth's staggering convulsion. To them Good Friday was every bit as traumatic as anything that has happened in L.A. or San Francisco. Hundreds of homes were damaged or destroyed in communities all along the south coast. In several water-front towns large oil storage tanks ruptured and caught fire, spreading a blanket of flames across the tops of incoming tsunami waves.

In Anchorage, a multistory apartment block and a big department store collapsed. Yawning cracks opened in downtown streets and slabs of falling concrete crushed cars like pop cans. Outside the city, in a neighborhood called Turnagain Heights, built on a clay bluff with spectacular views of the sea, the earth split open in dozens of places. The ground slumped, houses caved in, and people fell into a maze of crushed timbers and fissures that opened and closed in the liquefied clay.

As the land heaved and bucked, railway tracks got twisted and humped, highways cracked, and bridges were yanked apart. Docks and port facilities in Whittier, Cordova, and Homer were smashed by incoming swells. Undersea landslides created local tsunamis that struck tens of minutes before the main tsunamis arrived from offshore. A wall of seawater fifty feet high (15 m) slammed the vital seaport of Seward, road and rail gateway to Alaska's interior. Most of downtown Kodiak was inundated; the entire port city of Valdez was wrecked and would have to be relocated.

Later that night a radio operator aboard the oil tanker *Alaska Standard* tapped a frantic message in Morse code that said, "Seward is burning." Ham radio operators working from mobile units in their cars finally reached the outside world with fragmented reports of whole towns "wiped out by a great tsunami!" In the end, giant waves took the most lives while rock and mudslides and twisted, heaved, and fractured slabs of solid ground caused the greatest physical damage.

Alaska had always been a rough and tumble, seismically active kind of place, and longtime residents had grown used to the almost constant rumblings of nature. But there had never been one quite like this. Even scientists who knew the most about the state's geology were puzzled because nobody could say for sure which fault had broken or how it had gone undetected for so many years.

In a series of urgent phone calls late that Friday night, officials at the U.S. Geological Survey decided they needed to know where and how the earth had fractured and how a fault could lie silent, almost as if it were dormant, while storing massive amounts of energy for perhaps hundreds of years before ripping apart in a megathrust earthquake. Famous faults like the San Andreas and many of its lesser known cousins are obviously moving—creeping and slipping and breaking—somewhere almost all the time. Like schoolyard bullies, you always know they're not far away and could cause trouble at any moment. But for a really big fault to do nothing in all of recorded history and then suddenly rip itself apart—that was a mystery that had to be solved as quickly as possible.

Something very much like this had happened off the coast of Chile only four years earlier, in 1960. That event too was a mystery because scientists still did not know what had caused it—the biggest temblor of all. But first, and most immediately, they had to find out exactly what had happened in Prince William Sound.

CHAPTER 4

Against the Wind of Convention: Plafker, Benioff, and Press

On Saturday morning, March 28, 1964, thirty-five-year-old geologist George Plafker made a hasty exit from the conference in Seattle to join a team of USGS scientists dispatched to Alaska for a rapid investigation of the earthquake. He and colleague Arthur Grantz from the main West Coast laboratory in Menlo Park, California, wanted to see first-hand what had happened to the landscape—everything from rock and mudslides to compaction of the ground, the liquefaction of soils, the heaving up or dropping down of sections of land, and the effects of tsunami waves—all before anything changed or was cleaned up. They were also hoping to find the unknown fault that had torn the land apart and clues that might explain "the mechanism" of the rupture—how and why it happened.

Local officials in Alaska and U.S. military officers involved in the rescue and recovery effort wanted a quick damage survey that would warn them of any unfinished landslides, avalanches, or rivers that had been dammed by landslides—anything that might cause still more havoc and destruction. What they wanted to know more than anything else was whether the shaking was over yet. And if not, when might it start again?

Within twenty-four hours of the main shock, there had already been at least ten aftershocks of magnitude 6, along with dozens of smaller but nonetheless nerve-rattling vibrations. Seismometers would eventually log twelve thousand aftershocks in the first sixty-nine days after Good Friday. But any reputable scientist knows there is no simple answer to the question of when the *next* tremor might come. In this case they couldn't even say for sure what had caused the main event.

Plafker saw some of the worst devastation right away in the bluffs around Anchorage, along Turnagain Arm and near the harbor at Fish Creek. The land underneath had liquefied, causing the surface to fracture into blocks. "There were these large areas along the bluffs that had just kind of slipped seaward," he recalled. "They'd broken up—the blocks at the surface broke up and tilted. And there were several people killed . . . It was pretty overwhelming."

At the very least, Plafker and his colleagues figured they would be able to provide a detailed description of what the temblor had done to the surface of the land and which way the fault had come unstuck—assuming they could find it. What started out as a one-week reconnaissance mission to determine the scale of the thing turned into a summer-long operation requiring a whole team of USGS personnel to catalog and research the devastating aftereffects.

They hitched rides on airplanes and helicopters borrowed from the military and flew hundreds of miles along the coast, making detailed descriptions of a torn and ravaged land, searching all the time for evidence of a major fault. The shaking had triggered a series of long, jagged rockslides that left gaping scars on the slopes of heavily forested mountains. There were open fissures and heaved-up slabs of rock in several segments along these smaller escarpments. But there was no sign of a much bigger fault—no continuous rupture in the earth's crust, nothing obvious enough from the air to have caused an earthquake this widespread and violent.

Hundreds of seismometers around the world had recorded the

shockwaves from Prince William Sound, and early calculations pointed searchers toward a small glacial fjord in the mountain shoreline called Unakwik Inlet. Somewhere twelve to thirty miles (20–50 km) underground was the "focus," or starting point of the main rupture. Directly above the focus, where a vertical line would meet the surface of the earth, was the epicenter. But the epicenter was a problem; there was nothing to see, no physical damage to the surface that common sense would tell you ought to be there after one of the biggest earthquakes in the world had just broken the place apart.

Plafker's first impression was that the rupture zone lay hidden beneath new snow or was perhaps somewhere offshore. Several new science papers resurrecting the old theory of continental drift had been published only two years earlier and were still the source of controversy and buzz, but the exciting idea that a slab of the Pacific Ocean floor might be sliding *underneath* the state of Alaska might explain why the fault was not visible from the surface.

As the investigation continued, Plafker hiked the shoreline to take detailed measurements of vertical changes in the level of the land. What he discovered was truly astonishing. A segment of the earth's crust 430 to 500 miles long and 90 to 125 miles wide (700–800 km by 145–200 km) had been "deformed" by the earthquake. An area roughly the size of Washington and Oregon combined had been either heaved up or dropped down—"larger than any such area known to be associated with a single earthquake in historic times," he later wrote.

Somewhere between 66,000 and 77,000 square miles (170,000–200,000 km^2) of the ocean floor had been hoisted up, while vast areas of dry ground inland (*behind* the beach zones) had sunk. The sea floor southwest of Montague Island appeared to have been lifted more than fifty feet (15 m). The sudden upthrust of the ocean bottom was clearly what had displaced so much seawater and created the deadly tsunamis that hit Port Alberni and the West Coast.

Wearing gumboots and hauling a surveyor's level across the slippery rocks, Plafker spent most of that summer in Alaska making more than eight hundred separate measurements of uplift or subsidence (relative to sea level) along thousands of miles of shoreline between Bering Glacier and the Kodiak Islands. In some places he didn't need equipment to see what had happened. He could tell how far a beach had been raised simply by examining the whitish band of dead barnacles, algae, and mussels that had been killed when seafloor rocks were lifted above the reach of tides. Without their daily slosh from the ocean, all the sea creatures clinging to those heaved-up rocks had died and been bleached by the sun. Their reeking bodies painted a marker line on the rocks that measured how much the earth had moved.

In other places, where the ground had *subsided*, he found bands of dead brush that had been killed by seawater. As the incoming tides extended their reach over newly sunken beaches and marshes, the salt was slowly poisoning huge shoreline trees that once had lived above the tides altogether. It would take a while for the big trees along the beach to die and wither, but their mossy hulks standing in knee-deep, newly created saltwater lagoons would become vital clues in a later investigation.

Plafker found hoisted sea cliffs, drained lagoons, new reefs and islands—all indications of violent and widespread upheaval. A short time later another crucial piece of the puzzle came from the USGS survey crew, who rechecked a network of triangulation points and discovered that the earth's surface had also been *stretched* horizontally as much as sixty-four feet (20 m) between Anchorage and the outer island of Prince William Sound. This extraordinary piece of geographic distortion would eventually help prove what kind of rupture this really was and why nobody could see the fault from the surface.

Plotting his elevation numbers on a map, Plafker also noticed an invisible line of "zero change" in the level of land, approximately parallel to the south coast mountain ranges and to the deep Aleutian Trench offshore. On the seaward side of this invisible line, the land had been raised; on the landward side it had dropped down. "What I was doing was just trying to get some feeling for whether these areas of uplifted and subsided ground might be pointing to a fault in between them," he told me.

If there was a hidden crack in the earth, it seemed odd that heaving up and dropping down—especially on a scale as grand as this—could have happened *without* breaking the surface. How could so much land be jacked up or slumped with no visible fracture line? And yet "we never could see the fault," he said, and that made the Alaska mystery all the more fascinating.

In numerous places he saw the "squeezing up of the rocks," which he likened to a crumpled fender. It all looked very different from the kinds of surface damage he'd seen when plates slid past each other along a fault like the San Andreas. In California the earth was fractured vertically—and it was plain to see—but in Alaska the rocks were being folded up and shortened. Or stretched horizontally like taffy.

The essential unknown of the Good Friday rupture—the true nature of the fault—needed an explanation, so Plafker and his colleagues spent months living on a converted river tugboat, prowling the shore in small skiffs, measuring rocks and crunching numbers trying to make sense of what they'd found. Their data logs were so chock-full of bewildering new information it would take until June the following year to get it organized and published.

To make the job more challenging, Plafker, a relatively young scientist who had not yet earned his PhD, was preparing a report about earthquakes—not his chosen specialty. He was a geologist who'd spent most of his career up to that point mapping rock formations, searching for oil and other natural resources. He was not trained as a seismologist, yet here he was writing about an unseen fault that had behaved contrary to what most experts in the field were familiar with. This invisible crack along the Alaska coast appeared so unlike the San Andreas that the facts and figures Plafker came back with beggared belief. And got him into a bit of hot water.

∞

The new science that would eventually explain what happened in Alaska—the revolution in geology now known as plate tectonics—was in mid-evolution in 1964. The controversial theories had not been refined, tested, or accepted. "They were still just barely getting to it at the time of the earthquake," recalled Plafker. Strange as it may seem today, there was no broad consensus then on how mountains and volcanoes were formed or what kinds of forces generated earth tremors. Geophysicists didn't even know for sure whether faults caused earthquakes or, the other way around, earthquakes caused faults. Was the earth's surface cooling and shrinking and cracking? Or was it expanding and cracking because of radioactive heat from the deep interior of the planet? All these big ideas were still very much in play.

When I phoned him in 2009 to talk about the turmoil of the times, it was hard for Plafker to remember after so many years exactly what he knew when he flew north from Seattle that day in 1964, but one name did stand out. Hugo Benioff, who in the 1930s had designed and built the most sensitive earthquake detection equipment in use, was one of the three wise men who pioneered the young science of seismology at the California Institute of Technology (Caltech) in Pasadena. Benioff had written a classic series of papers between 1949 and 1954 that drew the first hazy picture of big slabs of the ocean floor thrusting underneath the margins of the Pacific Rim.

Benioff borrowed the voluminous and detailed charts of worldwide seismic data compiled by his famous Caltech colleagues, Charles Richter and Beno Gutenberg, to compile the first truly quantitative description of an earthquake mechanism. When he plotted on a map where most of the Pacific Rim ruptures had happened, Benioff noticed that they were not randomly distributed but instead concentrated in distinct zones of intense seismic activity parallel to most of the major island arcs, "curvilinear mountain ranges," and deep ocean trenches along the coastlines of the Pacific Ocean basin. He had charted in precise numeric detail the infamous Ring of Fire, as we know it today.

Benioff's 1949 study revealed the existence of two great faults, previously undiscovered: one off the coast of Tonga nearly 1,550 miles (2,500 km) long, the other off the coast of South America nearly 2,800 miles (4,500 km) long, both roughly 560 miles (900 km) wide and extending approximately 400 miles (650 km) downward into the interior of the earth. When he wrote that the South American sub-sea fault was "larger than any previously known active fault," it almost sounded like bragging. Eleven years later, when the fault ripped apart and wrecked the coast of Chile in the largest earthquake ever recorded on scientific instruments, his words turned out to have been prophetic.

Benioff explained, "The oceanic deeps associated with these faults are surface expressions of the downwarping of their oceanic blocks. The upwarping of their continental blocks have produced islands in the Tonga-Kermadec region and the Andes Mountains in South America." When he noted that "the continental mass flowed over the oceanic mass," it sounded like an endorsement of the still heretical theory of continental drift.

What Benioff observed was that blocks of continental land seemed to be thrusting up and sliding over top of blocks of the sea floor. Another way to see it might be that slabs of the ocean floor were diving underneath the continental coastlines, cutting deep trenches offshore, scraping rock against rock, generating volcanoes, building huge mountain ranges like the crumpled fenders of massive collisions—and causing earthquakes in the process. When these oceanic cracks or faults occur close to the edge of a continent, he explained, the seafloor slab extends downward at a shallow angle of roughly thirty-three degrees. Of specific interest to young George Plafker was Benioff's calculation that a fault under the Aleutian Island chain off the coast of Alaska dipped at an angle of twenty-eight degrees beneath the mainland.

The picture Benioff saw in a cluster of seismic dots—what geophysicists now refer to as a *subduction zone*—was a crucial missing piece of the still incomplete great tectonic puzzle. At the time he wrote, in 1954, a significant number of Benioff's fellow seismologists were unwilling to embrace a concept that ran against conventional wisdom. To most experts of the day, a fault was a nearly vertical crack in the earth. Ten years later though, as Plafker stood in his muddy boots on the wrecked Alaskan shore, Benioff's idea had the ring of truth.

As he began to write the first draft of his own report, Plafker looked at the plots of seismic data from Prince William Sound and concluded that a "low-angle fault" had caused the catastrophic earthquake of 1964. He described a crack in the crust that was almost *horizontal* instead of vertical—sideways compared to the San Andreas. He described a colossal continental collision in matter-of-fact terms, with what seemed to Plafker a logical conclusion about what had caused the rupture and why there was no visible fault at the surface. But a politically significant part of the science community did not agree.

"No, no, no. Not much of this was accepted," Plafker laughed. "Hell, people gave me big arguments about Alaska. And the trouble was that *one* of them happened to be a world-class scientist who later became the president's science advisor and the head of the National Academy of Sciences."

It came down to a question of geometry. Frank Press, the prominent seismologist at Caltech who later became science advisor to U.S. president Jimmy Carter, thought Plafker was wrong. He examined the seismograms from Good Friday, calculated what is known as a "fault plane solution," and concluded that "a near-vertical fault plane is uniquely indicated." How could Press and Plafker look at the same raw data and see such radically different pictures? Unfortunately, a fault plane solution does not provide a single, unambiguous answer.

The math will yield two potential answers, only one of which can be correct. One solution will be a geometric plane, a cross-section of the earth, that passes through the focal point of the earthquake. The *alternative* solution—another slice or cross-section that also passes through the quake's focal point—is perpendicular to the first. The angle of the fault (and thus the movement of rock slabs during the quake) will be along one of these two planes. But with no obvious rupture at the surface—with no way to examine the crack and "ground truth" the answer—these mathematical plots of shockwave data from deep underground become hypothetical. And in this case, controversial.

When he examined the diffuse pattern of the thousands of aftershocks, Plafker thought the correct solution was obvious: "If you have a big blur of aftershocks, like in '64, it seemed to me that you could say, 'Well, it's the low-angle plane because [the aftershocks are] spread out over a broad region.'" A vertical fault would presumably have produced a vertical or linear pattern of aftershocks. This one did not.

Indeed, the Alaska quake had generated an angular haze of twelve thousand small to moderate tremors that extended 90 to 125 miles (145 to 200 km) from the epicenter—the rupture began underneath the continental mainland—out to sea, reaching the inner wall of the deep Aleutian Trench. Plotted across a wide band of geography, these seismic dots looked like a bad rash on the underbelly of the continent. Connect the dots from the epicenter out to the trench and you might find the missing fault.

In June 1965, when his own paper was published, Plafker openly discussed the conflicting fault plane solutions and acknowledged the elephant in the room. He wrote that everything scientists knew about the angle of this invisible rupture zone had been "deduced from seismological data," the implication being that nobody could say for sure what kind of fault it was. "Neither the orientation nor the sense of movement on the primary fault is known with certainty," he wrote. They were all making educated guesses.

Three weeks later, Plafker's dissenting view made headline news. Walter Sullivan of the *New York Times*, a regular reader of the latest in *Science*, saw Plafker's paper and quickly cranked out a feature article that highlighted the split between Plafker and Press. "Data collected by a number of field parties during the last year has given birth to two alternative accounts of the mighty rupture within the earth," wrote Sullivan as he zeroed in on Plafker's main conclusion. "He believes that a mass of material from beneath the ocean floor suddenly thrust inland under the continental rocks." While most non-scientists probably missed the subtle implication, geophysicists around the world knew that Plafker had essentially said that the emperor had no clothes.

CHAPTER 5
Cauldron and Crust: The Rehabilitation of Continental Drift

A non-geologist might wonder why it really mattered whether the Alaska fault was vertical or horizontal, but the implications for the West Coast were dire. If Plafker and Benioff were right about that slab of ocean floor poking down underneath the continent, then in all likelihood the same would be true for British Columbia, Washington, Oregon, and northern California. An earthquake like the Good Friday disaster would affect millions of people if it happened to the Pacific Northwest. But the debate was by no means resolved by Plafker's research even though in hindsight the evidence appeared overwhelming and obvious.

The debate about faults and earthquakes and how mountains were built had been raging for five decades. In January 1912, Berlin-born meteorologist Alfred Wegener dared to express for the first time his idea of "continental displacement." Like others before him, Wegener had noticed that the coastlines of Africa and South America seemed to fit together as if they were pieces of a jigsaw puzzle and that parts of North America and Europe also seemed to match up. He speculated that all these far-flung land masses might once have been part of a single supercontinent (he named it Pangaea) approximately 200 million years ago.

His idea was that Pangaea had broken apart and that huge masses of land had moved sideways across the surface of the earth, jostling and crashing into each other over millions of years. While others had speculated about the significance of matching coastlines, Wegener had the audacity to put it in writing, making himself an instant lightning rod. Published in 1915, his book *The Origin of Continents and Oceans* triggered a controversy that was still raging in 1964. If Plafker and Benioff turned out to be right about how that fault in Alaska worked, then the essence of Wegener's big idea might be right as well.

Wegener's notion that "tidal friction" and differences in gravity caused by the earth's imperfect, oblate shape had caused the continental breakup and that huge slabs of the earth's crust somehow plowed across the ocean floors like ships through pack ice was considered by physicists to be impossible. What on earth—what mechanism—could possibly generate a force strong enough to fracture and move entire continents horizontally? To many it sounded like utter nonsense. The evidence that big land masses had once been joined together, however, was harder to dismiss.

The dominant view among scientists at the time was that the continents had been locked rigidly in place from the very beginning of time as the earth solidified from a molten state. With the interior of the planet gradually cooling, the outer crust began to shrink and slump, to crack and wrinkle like a drying apple's skin—creating mountains along the way. Others thought parts of the crust rose and fell periodically as if they were floating on a semi-fluid interior.

Most who doubted Wegener were aware of the work of other scientists pointing to fossil match-ups, the remarkable similarities between rock layers on different continents, and the evolution of nearly identical plants and animals on opposite sides of the oceans. They realized that sooner or later there would have to be some way to account for all this.

Wegener offered a theory to explain how the various bits and pieces *might* have fit together, even if he couldn't say for sure why they had come apart.

He argued that if continents could move downward as the planet cooled and contracted, or even upward—rising slowly as ice ages ended and the enormous weight of glaciers melted away—then they could probably move horizontally as well. Figuring out how and why this happened would be the crucial next step. Even before Wegener published his theory there were reasons to question the orthodox view that the earth was cooling and shrinking. In fact, some researchers already thought the opposite might be true.

With the discovery of nuclear radiation at the turn of the century came the understanding that some elements generated energy all by themselves, that rocks containing these elements deep underground might be pumping out an enormous amount of heat that accumulates faster than it can dissipate into space. If true, then perhaps the earth was heating up rather than cooling. The surface of the planet might actually be expanding rather than shrinking. It might also explain how continents could slide or drift sideways.

Scientists began to speculate that heat generated in the earth's interior might get trapped beneath the continents. Radioactive elements could be generating enormous "convection currents" of melted rock that would rise from the planet's white-hot mantle toward the surface, like bubbles in a pot of soup. In fact the soup analogy seemed to make so much sense it was still being taught in Geology 101 courses when I entered university in 1970.

I can still recall the lecture. The professor, whose name is lost to me now, asked us to imagine the earth as a large cauldron of thick soup that has been brought to a slow boil. Bubbles of heat rise up from the bottom of the cauldron. At the surface the soup cools and forms a crust that floats atop the hotter liquid material below. When new heat bubbles rise to the surface, they push the older crust aside.

Propelled against the outer walls of the pot, the older crust is pulled down into the interior of the cauldron, where it gets reheated and eventually bubbles back to the surface to form crust again. This continuous, circular motion of a heated liquid is now known as a convection cell. And that, concluded the professor, is how we might explain the way continents get dragged or pushed across the surface of the earth.

Just think of continents as great rafts of floating soup crust. In the 1920s, however, this was still just a wild idea with no solid evidence to back it up. Wegener and his supporters might have been cheered by these new developments, as they seemed to provide an explanation—the mysterious *force* that could move continents around like pieces of a jigsaw puzzle—and make sense of continental drift. Sadly that's not how the story ended for Wegener himself. A meteorologist by profession, he got lost in a blizzard during a research trip to Greenland in 1930 and did not live to see the discoveries that would rehabilitate his theory more than three decades later.

If continents were indeed moving around like rafts of soup crust, there had to be some way to prove it once and for all. The next several breakthroughs in earth science came as a result of military research begun during World War II. The U.S. Navy needed a new technology to detect German U-boats and better maps of the ocean floor to keep track of where enemy (and their own) submarines might be able to hide. In those days the bottom of the sea was as uncharted as outer space and a generation of young scientists was eager, willing, and able to explore the planet's last frontier.

As warships sailed from one battle to the next, echo sounders pinged day and night, creating detailed, never-before-seen profiles of the ocean floor. In the Pacific, they charted undersea volcanoes and steep canyons like the Marianas Trench which, according to the new measurements, was seven miles (11 km) deep. What process had created such a steep canyon in a mostly flat ocean floor? Could this be where the soup crust buckled under and got recycled into the earth's interior cauldron?

The threat of enemy subs seemed just as real during the Cold War, so the Office of Naval Research continued sending exploration teams to sea during the 1950s and early '60s. Along the way scientists discovered amazing new details about the Mid-Atlantic Ridge, a chain of undersea mountains halfway between Europe and North America. The so-called ridge turned out to be a set of parallel mountain ranges with volcanic vents oozing hot magma onto the ocean floor.

As the magma spewed out and cooled in seawater, it expanded and hardened to a rocky crust, forming a new piece of ocean floor. Over millions and millions of years, the lava had piled up into those volcanic ridges while a seemingly constant spew of new magma kept pushing the ridge flanks farther and farther apart. Here, at last, was direct physical evidence of the convection currents that might be causing continental drift.

At first it was thought this mid-ocean ridge existed only in the North Atlantic. Further mapping confirmed that it wandered down between Africa and South America and then snaked around the entire globe like the seam on a baseball, a fifty-thousand mile (80,000 km) chain of volcanic ridges. When they put all the new charts together, these mid-ocean ridges turned out to be the longest continuous mountain range in the world. In terms of scientific significance, the undersea ridges had morphed into the most prominent geologic structure on the planet. Research done on these volcanic slag heaps would make or break the theory of continental drift.

In the early 1960s, with hot convection cells to power the system, the next question to answer was what happens when two moving portions of crust collide. Like two cars crashing head on, the obvious result of two continents slamming into each other would seem to be crumpled fenders—mountain ranges like the Himalayas and the Alps. When a segment of ocean floor crashes against a continent, the sea floor, being made of heavier, denser, volcanic rock, apparently buckles under the lighter continental crust, creating a deep ocean trench like the Marianas.

As the heavier ocean floor continues to move, it gets forced downward, grinding against the underside of the continent as it goes. Coastal mountain ranges get shoved upward in the process. As the seafloor slab goes deeper, it gets so hot it begins to melt, spewing a volcano up through the overlying continental rock. If two masses get stuck together by friction instead of sliding past each other smoothly, enormous pressure builds up and is finally released in megathrust earthquakes. So the old soup cauldron story had endured, and the conversion of many skeptics into tentative believers was underway. Finally there was a logic to continental drift and a way to put the planet's jigsaw puzzle together.

By the mid-1960s J. Tuzo Wilson at the University of Toronto would weave the bits and pieces of new discovery together in a comprehensive theory that filled in most of the blanks in Wegener's original concept. Wilson also changed the terminology. He described the earth's surface as being divided into "several large rigid plates" rather than continents. By Wilson's definition, a plate was considerably larger than a continent and could include segments of ocean floor that had been jammed against and welded to the edge of a continent. The North America plate, for example, included everything from the Mid-Atlantic Ridge to the California coast—meaning all that "new ocean floor" being generated underneath the Atlantic had become part of the older continental mass. It was all of a piece, a westward-moving tectonic plate.

His choice of the word *plates* would allow a new generation of researchers to put the old bogey man of continental drift behind them and move forward into the emerging world of *plate tectonics*—the new geology. It was a great time to be an earth scientist because there was still so much to figure out about how the system worked, especially along the western coast of North America and around the Pacific Rim.

New voyages of discovery revealed that the mid-ocean ridge system—that twisty baseball seam of volcanic mountains circling the globe—cut through the wide, seafloor prairie of the Pacific Ocean, fracturing the main plate and pushing smaller pieces off to either side as it spread the sea floor wider. The Cocos plate, for example, had apparently been split off from the larger Pacific plate by a convection cell pushing new magma up through the East Pacific Rise: the segment of the baseball seam running parallel to the coast of South America. Upwelling magma had pushed the smaller Cocos plate eastward underneath Central America.

Farther south, researchers learned that another broken slab of sea floor was being thrust under the coast of Chile. To the north another was punching its way down beneath the beaches of Alaska. The same was happening under the coast of Japan and in many other places around the Pacific Rim—all because of seafloor spreading.

Anywhere you looked, broken plates were pushing against one another. At each one of these collision points were large mountain ranges, violent earthquakes, and active volcanoes. The Pacific was circled by a "ring of fire" caused by lumps of the earth's crust crashing together, melting and erupting.

While scientists around the world were busy piecing it all together in their minds and on paper, the earth itself was providing physical proof of what was really at stake. In 1960, the broken chunk of ocean crust jammed beneath Chile's continental shelf finally reached its breaking point and snapped loose in the largest earthquake ever recorded. Scarcely four years later, George Plafker was collecting evidence that the same kind of horizontal fault had caused the 1964 earthquake in Alaska. Even though the old guard had still not accepted the idea of plate tectonics, Plafker was pretty sure he was right.

One could argue that this should have been the dawning of awareness of the megathrust earthquake threat to British Columbia, the Pacific Northwest, and California as well. A 1965 paper by Tuzo Wilson pointed to the existence of what he called the Juan de Fuca Ridge. The name was chosen because the upper end of the ridge lay due west of the Strait of Juan de Fuca, which runs between Vancouver Island and Washington State. Here was an undersea mountain range that had previously been discovered and then dismissed as an insignificant, amorphous hump of rock running parallel to the coast. But if Wilson and the young turks of plate tectonics were right, the Juan de Fuca Ridge was in fact another part of that fiery seam of volcanic mountains running through the oceans.

If this ridge turned out to be spreading apart sideways, powered by a cauldron of hot magma, it must also be thrusting a slab of sea floor underneath the edge of British Columbia, Washington, Oregon, and California. Presumably some kind of trench would be located where the two plates met, a "convergent plate boundary" just like the ones off Chile, Alaska, and Mexico. If so, giant earthquakes must surely follow.

CHAPTER 6

Nuke on a Fault: Early Clues in Humboldt Bay

Flying south from the Oregon line in search of tectonic damage, the helicopter finally angled west over the last wall of mountains and down through a hole in patchy clouds to the California shore. We found the Shelter Cove runway, a cracking strip of sun-baked asphalt surrounded on three sides by a golf course on a bench of land just above the sea. My first thought was—how quintessentially West Coast. Fly in for a quick round of golf, go whale watching in the afternoon, have a barbecue on the beach, then fly home at sunset. Just try not to think about that monster earthquake hiding beneath the surf.

On this particular summer morning in 2007, cameraman Doug Trent and I had planned to shoot aerial pictures for a new documentary called *ShockWave* on the communities closest to the Juan de Fuca Ridge and fault: the farms, ranches, and small towns from Cape Mendocino north to Humboldt Bay, Eureka, Arcata, and Crescent City, near the Oregon border. There was a reasonable chance the morning fog would burn off by midday, giving us the low-angle light we needed to highlight a half-dozen surface-level fractures where ancient tectonic ruptures had heaved up beaches and hillsides along the foreshore.

Geology professor Lori Dengler at Humboldt State University had supplied Doug and me with a list of sites to photograph, complete with GPS coordinates that made them much easier to find. She referred to this section of California coast as a "fold and thrust belt," the crumpled edge of a tectonic subduction zone.

The first thing we photographed was a cleft in the hills directly behind Shelter Cove, the northernmost mapped trace of the San Andreas fault. Through the open door of a JetRanger it looked like just another deep shadow among the redwoods. It was bizarre to think this darkish line was in fact a crack in the earth's crust, the constantly creeping boundary zone between the Pacific and North American plates.

In the earthquake of 1906 the San Andreas tore itself apart to the north and south of San Francisco. The northern segment of the rupture ran 250 miles (400 km) from the Golden Gate all the way up the coast to Cape Mendocino—westernmost point of land in the lower 48 states—leaving this visible crack in the hills behind Shelter Cove. Even at such a distance from the ruined city, shockwaves here were strong enough to crack walls, break windows, and topple chimneys in the nearby farming towns of Ferndale and Eureka.

At Cape Mendocino itself, the San Andreas disappears offshore. In the aftermath of the San Francisco disaster, scientists began to speculate and disagree about where the fault goes from there. Does it continue north toward Alaska? Or does it veer west out to sea? The routing or northern extension of California's most famous and deadly fault remained a mystery for decades. By the mid-1960s, however, the emerging theory of plate tectonics seemed to promise a better understanding of how this amazingly complex system of cracks worked.

Geologists and oceanographers knew from mapping the underwater terrain (the bathymetry) that the ocean floor looked like a broken dinner plate. From what they could see with the earliest, relatively low-tech echo-sounding equipment, the bottom of the Pacific Ocean had cracked in several places. The recently "rediscovered" (and newly named) Juan de Fuca Ridge rose from the deeps off the northern California coast, fracturing the sea floor in a northwesterly direction toward Vancouver Island, more or less parallel to the coast.

A convection cell of hot magma from the earth's mantle had apparently broken the Pacific plate apart, shoving a slab of oceanic crust (the Juan de Fuca plate) east underneath the oncoming (westward-moving) North America plate. This would eventually become known as the Cascadia Subduction Zone. On closer examination, researchers discovered that the Juan de Fuca plate itself had been fractured. The southern end of it was broken off and appeared to be moving independently.

It turned out there was a separate, smaller ridge—another seafloor spreading zone called the Gorda Ridge—that looked like a southern extension of the Juan de Fuca system. It appeared to be pushing another chunk of oceanic real estate, the Gorda plate, beneath the California coast. There was also a heaved-up fracture zone running east to west across the larger Pacific plate. All these cracks and broken slabs, including the San Andreas, converged offshore at Cape Mendocino. Geologists decided to call this tectonic wreck the Mendocino Triple Junction.

The ongoing and extremely slow-motion convergence of plates had fractured, bent, and folded rocks along the shore and hoisted up the beaches in several places, creating terraces we could now see from the air at a place called Singley Flat. They were, on a smaller scale, similar to the sections of heaved-up coastline George Plafker had discovered in Alaska after the 1964 disaster. If I had not been told what to look for, I would never have guessed these grassed-over benches of coastal farmland were the bent fenders of a continental crash, evidence that Cascadia's fault had caused numerous earthquakes over the years.

Farther north we shot pictures of the ruptured earth at places called Little Salmon River, Mad River, and McKinleyville. These were even harder for our untrained eyes to notice because today they are camouflaged by a veneer of human civilization, the streets and homes, schoolyards and shopping malls of Eureka and Arcata, California. Who would notice that the nice little house on what looks like a landscaped terrace is in fact perched on the edge of an active, still-moving fault, a fractured wedge of crust that is being shoved upward by the force of plate tectonics?

Native people who have lived in beachside villages along this coast for thousands of years tell stories they learned from their elders of horrific ground shaking on a winter's night long ago, followed by a killer wave that wiped out entire communities. For the most part, though, the tide of white settlers who began arriving here in the 1850s to homestead and log the redwood forest were unaware of, or simply uninterested in, the local knowledge of Aboriginal people. After 1906, they knew—the whole world knew—about the deadly San Andreas, but the concept of plate tectonics and the fact that a moving block of ocean floor could cause an even larger shock was still unknown.

The scientists and engineers who in the early 1960s drew up plans for an atomic power plant—California's first commercial reactor—to be built on the shore of Humboldt Bay assumed there was no major seismic threat to worry about. Ironically, building the reactor would help scientists discover the reality of Cascadia's web of faults.

Around the corner and up the coast from Cape Mendocino, the side-by-side beach towns of Eureka and Arcata were inhabited in 1963 by an uneasy mix of loggers, commercial fishermen, and back-to-the-land idealists who would soon be labeled hippies and environmentalists. The biggest construction project in decades—the atomic power station at Humboldt Bay—was coming to an end and the Pacific Gas and Electric Company (PG&E) was about to deliver a sixty-thousand-kilowatt jolt to the local economy.

Even though terms like *meltdown* and *China syndrome* had not yet colored the vocabulary, a tide of negative opinion had already derailed another nuclear plant downstate, so the residents of Humboldt County were aware of the potential for controversy and had mixed feelings about the promise of "power too cheap to meter." Work at Bodega Head, about fifty miles (80 km) north of the Golden Gate, had been delayed by vigorous opposition for six years. Construction workers had managed to dig a deep hole for the foundations when geologists confirmed that the San Andreas fault ran right beside (some said directly underneath) the reactor site. Eventually, PG&E decided to abandon the project.

Perhaps because the San Andreas veered out to sea at Cape Mendocino, state officials agreed with PG&E that seismic risk would not be an issue on the north coast at Humboldt Bay. With no large earthquakes in the area—at least not since the 1850s, when white settlers started keeping written records of local history—project managers at the utility and engineers who were designing the reactor were convinced the level of risk was within acceptable limits.

The federal Atomic Energy Commission, which would eventually license the plant, defined an active fault as having had one "event" (earthquake) in the past 35,000 years. Or more than one in 500,000 years. AEC regulations in effect at the time specified reinforced, anti-rupture reactor vessels only when an active fault came within a quarter of a mile (0.4 km) of a plant. But with no detailed information available about the quake history of nearby faults, and with little understanding of the newly discovered Mendocino Triple Junction or the implications of plate tectonics, any seismic threat seemed distant and hypothetical. Not a problem at Humboldt Bay.

"The plans for it were being developed as Tuzo Wilson's first papers on plate tectonics were coming out," Lori Dengler told me. As a professor of geology and then department chair at Humboldt State University, she studied the plant's history and the simultaneous dawning of awareness of Cascadia's fault. "There was absolutely no inkling of a subduction zone or great earthquakes," she said. PG&E, trusting the best science available at the time, signed a contract to build the reactor six miles (10 km) south of Eureka.

After the showdown at Bodega Head, the Humboldt plant became a test case for PG&E. Company officials and state and federal politicians no doubt wanted to prove that atomic power could be harnessed safely. The reactor started boiling water and generating electricity for the northern California grid in August 1963. Less than a year later, however, two plates shifting on a fault in Alaska focused worldwide attention on tectonic theory and sparked intense debates about seismic risk on the West Coast. Indirectly this heightened awareness would ultimately shorten the lifespan of the power plant near Eureka.

Ironically, it was the tsunami rather than the earthquake that hit home first.

When the Good Friday earthquake of 1964 wrecked the south coast of Alaska, it sent a train of waves crashing down the coast all the way to California and into the streets of Crescent City, where more than a dozen people were killed. Crescent City is only an hour's drive north of the reactor on Humboldt Bay. Yet the vulnerability of the power plant to a similar wave or to a massive seismic rupture (or even to a local earthquake) did not occur to people living in the area. Not right away.

All eyes were on the immediate tragedy and its aftermath. The notion of an Alaska-size quake from a subduction zone just twenty or thirty miles (30–50 km) off the California coast struck no fear in the hearts of Eureka, Arcata, or the other small communities of northern California. Only those scientists who'd read—and were convinced by—the latest research realized what the Alaska story meant to the rest of the coast.

Many frontline scientists themselves were still at odds over what to make of plate tectonics. In February 1965, almost a year after the biggest known earthquake in North American history, Frank Press, the distinguished seismologist at Caltech, published a paper describing the fault that had wrecked the Alaska coast as a steep, vertical crack in the ground, similar to the San Andreas. Four months later George Plafker, the young geologist who did the muddy-boot work of measuring and documenting the aftermath, published his own paper saying just the opposite: that the fault ran at a shallow angle from the sea floor underneath the continental landmass. Plafker thus cast his vote in favor of tectonic theory while Press, the established authority, took a more cautious position—and his was still the majority view.

A nearly identical disagreement about the Mendocino Triple Junction arose three years later. A team of scientists at the University of California seismology lab in Berkeley released a paper in 1968 suggesting the San Andreas did indeed extend out to sea from Cape Mendocino and was in the process of cracking its way toward Alaska. Searching through seismograms for clusters of quakes, they discovered that Cape Mendocino was trembling at a rate "two to three times that of the combined central California clusters," which seemed to confirm that those broken slabs of sea floor were on the move, just as Tuzo Wilson had predicted they would be.

Bruce Bolt and his colleagues at Berkeley calculated a series of fault plane solutions and decided that many of the smaller tremors were generated by cracks running northwesterly *through* the Gorda plate toward Alaska—not *northeasterly*, as you might expect if the Gorda plate were being pushed underneath the California coast by seafloor spreading. Essentially Bolt and his team believed the region was "dominated tectonically by the San Andreas." No mention in their paper of spreading ridges offshore, low-angle faults, or pieces of oceanic crust getting stuck underneath the continent. The Alaska threat did not apply to California in their view.

Then along came Eli Silver from the U.S. Geological Survey in Menlo Park. In 1971 he wrote a new paper on the tectonics of the Mendocino Triple Junction and came to the opposite conclusion from Bolt. Recall that there are always two possible answers to a fault plane calculation—one being ninety degrees different from the other. While Bolt chose the northwesterly solution, Silver was sure the northeastern vector (and tectonic theory) was correct.

Meantime, as if anyone needed a reminder, the San Andreas showed once again that it still had the power to shock, wreak havoc, and dominate the attention of scientists and the general public. On February 9, 1971, the magnitude 6.6 Sylmar earthquake struck the San Fernando Valley, killing sixty-five people and rattling Los Angeles emergency planners, police, and firefighters like nothing had in years. Seeing two hospitals (one of them brand new) and a dozen freeway bridges collapse prompted state politicians to pass a new law (the Alquist-Priolo Earthquake Fault Zoning Act) that would tighten building codes and severely restrict construction of residential buildings and critical infrastructure on or near any active fracture zone in California.

Only a few years later, two geologists hiking the north woods of the California coast would quietly shift the focus back to the atomic power plant at Eureka. Gary Carver, a freshly minted professor at nearby Humboldt State University, and Tom Stephens, one of his senior thesis students, were conducting field research on a forest company's land in the upper Mad River drainage basin east of Arcata when they discovered some "very large and previously unknown and unmapped faults."

The Simpson Timber Company had kept hundreds of miles of access roads closed to the public during the 1950s, '60s, and '70s while they logged off huge stands of old-growth redwood forest. None of the land had been geologically mapped and Carver and Stephens didn't quite know what to expect. "We were able to go back in where geologists hadn't been for a very long time," Carver told me.

Now, with the big trees gone, the ground was nearly naked and rock formations were easier to see. Still, Carver and Stephens had to hike for miles and miles along steep switchbacks, mapping and following a web of fractures from the mountains all the way west and downhill to the intertidal and beach zones along the coast. They noticed a distinctive angularity, what they called a "rhombohedral fracture" pattern, which to a non-geologist's untrained eye would look like nothing more than "tiny little cracks in the sand" of cutbanks sliced through the wilderness by road builders for the logging crews.

This rhombohedral pattern, Carver explained, was how faults propagate through unconsolidated sand deposits. This was loosely packed sand left behind when this part of the coast was under water, a wedge of ocean sediment that had been shoved against the continent and now stood well above the high-tide line. "Instead of a nice, clean, one-plane fault in which two pieces of the earth's crust move past each other, it becomes hundreds or thousands of little tiny faults all closely spaced together," he elaborated. "I thought this was really neat."

Each of the fractures Carver and Stephens found cut through geologically young terrain, suggesting the cracks were relatively recent. That meant whatever tectonic force had caused the fractures might still be an ongoing threat. Carver was pretty sure the rhombohedral fractures had been caused by plate convergence and compression. While none of the individual cracks had a huge amount of movement, taken as a whole the offset was significant.

"We realized that these little tiny fractures we were seeing in many places were parts of faults that had very large amounts of displacement on them," said Carver. The displacement added up to several miles in total. "And again—you're sittin' right there on the edge of the mapped subduction zone and you see those big folds in young sediment," said Carver, "and you can't help but think that that subduction zone is still active." All of this within a few miles of the nuclear plant at Humboldt Bay.

That's when Carver decided to fly north to Alaska for a first-hand look at what had happened there. He needed to "see what a big earthquake looked like" in all its mangled glory so that he could better understand what he was seeing on the ground in California.

CHAPTER 7

Proving the Doubters Wrong: The Chile Connection

Gary Carver spent an entire summer in Alaska looking at the aftermath of subduction. He flew the entire length of the '64 rupture, every mile of broken shoreline. He also met George Plafker, who was more convinced than ever that the primary fault that had caused the beaches and bays to heave, buckle, and subside could not have been vertical. Plafker had recently returned from Chile and was eager to tell anyone who'd listen that the two biggest earthquakes in recorded history had caused exactly the same kinds of physical damage to the landscape.

Not only that, but some very prominent senior scientists were apparently coming around to Plafker's point of view. Frank Press, who had so famously disagreed about the angle of the fault, sat in the audience at the 1968 meeting of the American Geophysical Union in Washington, DC, and listened to Plafker's presentation of his paper on Alaska. The main theme of that year's convention was "The New Plate Tectonics," and here was Plafker telling the science establishment that the Alaska quake had been caused by two huge slices of the earth's crust converging almost horizontally, getting stuck together, and then snapping apart.

Plafker told me that after the speech Press cornered him and unloaded. "He came up and he was real mad," Plafker recalled. "He said, 'You know, I've written a lot papers and I've seldom been proven wrong. But you did it to me this time!' He told me I had caught him in the biggest mistake he made in his career," said Plafker. "His views on the mechanism of this earthquake had changed and he was man enough to say so."

But not everyone was convinced. Clarence Allen, another senior scientist who'd heard Plafker's talk, still needed convincing. And he threw down a challenge that Plafker simply could not resist. How, he asked, could this underthrusting of the ocean floor be happening only in Alaska? Did Plafker think that's what happened in the Chile earthquake as well? Plafker said yes—even though he didn't know for sure—and so Allen arranged the funding necessary to send him south.

Plafker spent two months scouring the Chilean coastline by car along the mainland and by chartered boat in the islands of the southern archipelago, measuring areas of heaved-up and down-dropped land. "There again in Chile, in the southern part, vegetation grows right down to the shorelines," he said. "You could see the effects of subsidence from the drowned and dead trees and brush." He found a zone of "tectonic warping, including both uplift and subsidence," that was 125 miles (200 km) wide and roughly 625 miles (1,000 km) long. It affected an area of at least 50,000 square miles (130,000 km²) in southern Chile.

The two-day series of temblors in 1960 had included two main shocks, thirty-three hours apart, along with fifty-six large aftershocks. The sequence of ruptures and the tsunamis they triggered killed 2,000 people in Chile and 230 more in Japan, Hawaii, and the Philippine Islands.

The main finding of Plafker's paper, however, refuted the previous conclusion that Chile's wreckage had been caused by a nearly vertical strike-slip fault (like the San Andreas) because it was based on "incorrect" data that were "clearly incompatible" with his newer evidence of tectonic movement. He wrote that the Chilean main shock "resulted from a complex rupture on a major thrust fault or zone of thrusting roughly 1,000 km long that dips at a moderate angle from the continental slope beneath the continental margin." In other words, the fault was more horizontal than it was vertical, just like in Alaska.

Plafker estimated that to cause such widespread upheaval and deformation of the landscape, there must have been at least 65 feet (20 m) and perhaps as much as 130 feet (40 m) of horizontal slip once the fault broke. This might seem "surprisingly large," he wrote, but "not excessive" if compared to the horizontal thrust of roughly 65 feet he had seen in the 1964 Alaska earthquake. The bottom line appeared to be that the events had both been caused by the same process: two pieces of the earth's crust crashing together.

So his gambit in Chile had been a success; he was able to prove plate convergence. "It was very straightforward, once you know what you're looking for," he said, putting the apparent mistakes of the first scientists on the scene in Chile into some kind of context, "but you know, it's just like anything else. If you've done it once before, it's a cinch. And if you haven't, you don't know what to do."

Nevertheless, when Plafker met Gary Carver a few years later and they compared notes on Alaska, Chile, and northern California, the similarities were hard to miss. "It's pretty clear to me," said Plafker, "that the southern end of Cascadia is very much like the eastern end of the Aleutian Arc and the area where the '64 earthquake occurred. We have the same type of continental margin."

When I asked Gary Carver why he thought it took so long for most geologists to come around to the view that Cascadia was a threat, he could remember clearly one paper—written by Masataka Ando of the U.S. Geological Survey and Emery Balazs of the National Geodetic Survey—that stood out. They believed the subduction zone had "foundered" and was no longer active. "They related the idea that the rise [the Juan de Fuca Ridge where the sea floor was spreading apart] was so close to the trench that the plate was too hot to go down," Carver explained. "It couldn't go down, so therefore, subduction had stalled." This would presumably explain why there had been no large subduction earthquakes in all of recorded history.

But Carver was now infected by Plafker's enthusiasm. He was sure those cracks in the sandstone meant something significant. Back at work in California he and Tom Stephens continued their research on the fracture zones along the northern California coast. They mapped each individual rupture and gave it a name—the Big Lagoon, Trinidad, McKinleyville, Mad River, and Fickle Hill faults. "As far as I know," said Carver, this was "the first recognition of the existence of large, active thrust faults north of the Mendocino Triple Junction." It was also the first onshore evidence of tectonic motion on the southern end of the Cascadia Subduction Zone.

The discovery of unknown crustal cracks on logging roads in the hills behind Arcata made geology a hot topic for students and other scientists working in the area. One of those drawn to a series of talks that Gary Carver gave in 1974 was Tom Collins, a geologist working for the U.S. Forest Service, based in Eureka at the Six Rivers National Forest office. When Collins saw slides of the "rhombohedral fractures" and heard Carver speculate about the relation between the faults in the hills and the big subduction zone offshore, he wanted to find out more about it.

Collins went exploring on his own. He knew about the Little Salmon fault, which had been partially mapped back in 1953 by a local geologist named Bud Ogle, and perhaps because it was the one closest to where he worked in Eureka, Collins decided to have a closer look. Across Highway 101 from the nuclear power plant, he wandered into a recently excavated sand quarry at the base of Humboldt Hill. There he discovered, completely by accident, more of those rhombohedral fractures that Gary Carver had talked about in his lectures.

A day or two later Collins phoned Carver who agreed to join him at the sand pit for a quick recon. "We recognized the Little Salmon fault extended further north than Ogle had mapped," said Carver. Here again "young material" had been torn, meaning the shockwaves that had caused those distinctive fractures had occurred not so long ago in geological time. Even more worrisome, it looked like the crack probably continued right underneath the highway and onto the 143-acre (58 ha) site where PG&E had built the Humboldt Bay reactor.

So Collins wrote up his discovery and, as a concerned citizen, sent it to the Nuclear Regulatory Commission (the new name for the Atomic Energy Commission) in Washington. It was the first in a long and increasingly political chain of events that galvanized local antinuclear activists who had formed the Redwood Alliance to do battle with PG&E. It also, as an unintended consequence, accelerated the scientific research that would finally confirm the true nature of the Cascadia Subduction Zone.

A magnitude 5.2 earthquake shook the town of Ferndale on June 7, 1975, causing repeat damage to a town that had barely survived the pounding of 1906. The shockwaves also hit Humboldt Bay to the north of Ferndale, and in the aftermath fresh cracks were discovered in the concrete pavement of the road leading into the nuclear reactor site. A team of engineers from the University of California at Berkeley was called out to study the "ground motions and structural response" at the power station. The concrete caisson, with walls four feet (1.3 m) thick and an outside diameter of 60 feet (18.3 m), dug 85 feet (26 m) into the ground, appeared to be okay. But PG&E decided to err on the side of caution and ordered a thorough examination just in case.

The Berkeley report confirmed that there had been no significant damage to the reactor. The summary page, however, spoke volumes. "The regulatory requirements led to an *adequate but not excessively conservative* margin of safety based on the motions recorded in this event" (my emphasis). In other words, PG&E had followed all the rules and nothing bad had happened this time, but if there was any chance of larger earthquakes, then all bets were off.

Roughly a year later, in July 1976, when the reactor was shut down for routine refueling, the seismic safety questions were red-flagged by the Nuclear Regulatory Commission. The NRC decided to keep the plant closed until the Little Salmon fault and the new system of fractures discovered by Gary Carver and Tom Stephens could be checked and the seismic hazard issues dealt with.

PG&E hired several consulting firms to conduct field studies to find out whether any of the faults were still active. A sixteen-station array of seismographs was installed in the surrounding mountains and along the northern California coast to get a more detailed picture of all the tectonic motion. In addition, the NRC decided to send its own team of scientists into the field to follow up on the work done by Carver and Stephens.

They created a timeline of earthquakes in the region. With backhoes they dug trenches across the Little Salmon and Mad River faults for close-up looks at where and how often the various layers of soil and rock below ground had been torn apart. Taking samples of woody debris, dead plants, and the remains of tiny sea creatures contained in the layers disrupted by quakes, they used radiocarbon dating to figure out when the ruptures had happened.

In the fall of 1980 the geologists concluded that the Little Salmon fault was indeed active and that it probably ran underneath or very close beside the reactor. The bottom line according to Woodward-Clyde Consultants, hired by PG&E, was that the seismic issues could be dealt with but the job would be neither cheap nor easy.

At this point two other factors may have entered the equation for Pacific Gas and Electric. On March 28, 1979, while the Woodward-Clyde team was still documenting the gritty details of the Mad River area and how it might affect the reactor at Humboldt Bay, things went alarmingly wrong at a nuclear power station called Three Mile Island in Pennsylvania. A relief valve got stuck open, allowing large amounts of radioactive coolant to be released into the atmosphere. The reactor core overheated and barely survived a partial meltdown.

The accident, while not as catastrophic as it might have been, helped turn the tide of public opinion against nuclear power. By some masterstroke of luck or serendipity, a Hollywood movie called *The China Syndrome* had been released only twelve days before the Three Mile Island accident. The eerily prescient film became an instant box office hit and probably did much to seal the fate of nuclear power in the United States. After months of investigation and analysis, the Nuclear Regulatory Commission issued a new set of far more stringent safety rules that would apply to all reactors, including the one at Humboldt Bay.

Add to this the legal, political, and financial implications of California's own new seismic zoning law, the Alquist-Priolo Earthquake Fault Zoning Act, which was passed in the aftermath of the Sylmar temblor, and the job of retrofitting and upgrading the reactor at Humboldt Bay became too expensive to be economically feasible for PG&E. Four years later the utility applied for permission to decommission the reactor permanently.

In the aftermath, an official report to the U.S. Geological Survey described the twenty-five-mile (40 km) Little Salmon fault as "part of a broad, compressional fold and thrust belt developed in the accretionary wedge above the Cascadia subduction zone." An accretionary wedge is formed by the sediment and pieces of seafloor crust piled up in a trench where two tectonic plates collide. Think of the North American continent drifting west like a snowplow across the sea floor, scraping up muck and compressing it into rock.

In most cases the wedge is found under water. From Vancouver Island all the way south to the Oregon–California border, this folded and buckled sedimentary wedge is piled up against the continental shelf dozens of miles offshore, where it's difficult and expensive for scientists to study. Only in northern California was it piled up right in plain sight and on dry ground. The towns of Eureka and Arcata were built on top of it, which is why Gary Carver and others at Humboldt State University were able to draw such a revealing picture of what Cascadia's fault was actually doing. They took advantage of a unique geological setting to make an important discovery.

If the Little Salmon fault was active, then the Gorda plate—which had caused the cracks—had to be active as well, pushing its way underneath California while North America plowed west. The subduction along Cascadia's fault had not "foundered," and the plates had not stopped moving. At least that was the conclusion I drew from reading the science papers and from interviewing both Plafker and Carver.

Taken as a whole, the Humboldt Bay power project had a significant but unintended consequence. Building a reactor on top of a crack in the crust—a crack directly related to the Cascadia Subduction Zone just offshore—generated the new science that provided the first physical evidence that the northern section of the California and Pacific Northwest coast faced the same kind of tectonic disaster as the ones that happened in Alaska and in Chile.

If I'd been a journalist in California back in the 1970s, I like to think I would have turned this story into headline news. But the immediate impact of these discoveries confirming continental drift was almost nil. The story of Cascadia's fault got lost in the controversy over nuclear power. Fortunately the scientists on the ground knew they were working on significant stuff and refused to quit.

It was the heady, meaningful kind of research that made it an exciting time to be a geologist—especially in the Pacific Northwest. Frank Press may have changed his mind, but many others in the science community still refused to buy the new geology. Even when the top half of a mountain in southern Washington State exploded, only a handful of researchers recognized the distinct sound of Cascadia's smoking gun.

CHAPTER 8
Mount St. Helens: Cascadia's Smoking Gun?

Even though geologists and volcanologists saw it coming, there was no way to prepare for the impact of watching a mountain explode at close range. Mount St. Helens—roughly ninety miles (145 km) south of Seattle and fifty miles (80 km) northeast of Portland—blew steam and dust for two months as a bulge of hot rock sprouted like a giant goiter on its north face. At the same time, the ground trembled and shook. People in downtown Portland turned the prelude into a spectator sport.

Government officials issued repeated warnings to evacuate the hills and valleys around the volcano as the frequency of tremors began to increase. Almost everybody did leave, except for an eighty-three-year-old recluse named Harry Truman who had lived in the woods near the mountain for more than fifty years and decided to stay close to his cabin. The media fell in love with him, a tragic hero in the making. A thirty-year-old volcanologist named David Johnston was collecting data until the very last minute. His final words, "Vancouver! Vancouver! This is it!" were shouted into a walkie-talkie and received at the USGS volcano observatory in Vancouver, Washington, across the Columbia River from Portland, only moments before the eruption. Neither man was seen again.

At 8:32 a.m. on Sunday, May 18, 1980, the volcano started coming apart. A magnitude 5.1 earthquake caused the bulging north side of the mountain to collapse where a new lava dome had been growing. The collapse caused the largest landslide of rock and ice and volcanic mud ever recorded in the continental United States—9,600,000 cubic yards (7,340,000 m³) of boulders, muck, trees, and other debris was swept 17 miles (27 km) downhill into the Columbia River. With the face of the mountain suddenly exposed to cool air, the volcano exploded, flattening or burying more than 230 square miles (595 km²) of forest and farmland under a blanket of mud and ash that shot 12 to 16 miles (19–26 km) into the sky.

Even though most residents had fled the area, the explosion still killed 57 people, destroyed or severely damaged more than 250 homes and businesses, wrecked 185 miles (298 km) of highway and 15 miles (25 km) of railway track, punched out 47 bridges, and killed more than 7,000 big game animals (deer, elk, and bear) and an estimated 12 million fish at a nearby hatchery.

Before the eruption Mount St. Helens had a spectacular, nearly symmetrical, cone-shaped peak that stood 9,677 feet (2,950 m) high—a stratovolcano. It collected 140 inches (356 cm) of rain and up to 16 feet (5 m) of snow every year, making it look a lot like those famous pictures of Mount Fuji in Japan. After the explosion, the top thousand feet of the mountain had vanished, leaving a horseshoe-shaped crater two miles wide and a half-mile deep (3.2 km by 0.8 km). As the eruption continued for nine hours, the ash plume drifted east at an estimated 60 miles (100 km) per hour, dumping a thick layer of abrasive grit across eastern Washington and Oregon, coating cars as far north as Edmonton, Alberta, as far east as the Dakotas, and as far southeast as Colorado and New Mexico.

The first warning signs had come as early as March 20, when a mild tremor (magnitude 4.2) rattled the mountain. Steam vents began to spew a week later. By the second week of April, scientists had alerted the media. Walter Sullivan of the *New York Times* touched on the explanation of Mount St. Helens' deep tectonic origin in his story "The West Is Alive with the Sound of Volcanoes."

The violence of volcanoes like this, according to the *Times* story, was a direct result of the collision of North America with the Pacific Ocean floor. It was the Juan de Fuca and Gorda plates grinding down along Cascadia's fault that had created the Cascade Arc of eighteen major volcanoes from Mount Shasta and Lassen Peak in California to Mount Adams and Mount St. Helens near Portland, to Mount Rainier near Seattle, Mount Baker, about fifteen miles (25 km) south of the Canada–U.S. border, and to Mount Garibaldi, north of Vancouver. Sullivan's feature explained what would probably happen—and why—a full month before Mount St. Helens blew: "The Cascades, part of the Pacific Ocean's necklace of volcanoes, its 'ring of fire,' are the product of 'sea floor subduction'. . . Typically, the sea bed bends down as it nears a continent, forming a trench. A sloping zone of earthquakes marks its path into the earth's interior. When the sea floor slab reaches a depth of about 75 miles [120 km], part of it apparently melts and, lighter in weight than the overlying material, forces its way up to produce volcanoes."

When I thought about the timing and content of Sullivan's story, it struck me that the eruption of Mount St. Helens should have been Cascadia's smoking gun—clear, unequivocal, physical evidence that continental drift was real, that the Gorda and Juan de Fuca plates were on the move and dangerous. This should have been the big wake-up call or tipping point for everyone involved in the geophysical sciences and emergency preparedness, a stark statement that the coast of northern California, the Pacific Northwest, and southwestern British Columbia were every bit as threatened by megathrust subduction disasters as were the coasts of Alaska and Chile. I was wrong. For reasons unclear to me, the alarm bells did not ring.

The Mount St. Helens disaster happened right in the middle of the investigation of those thrust faults that threatened the nuclear reactor at Humboldt Bay, yet it had no discernible impact on the reluctance of some scientists to accept the Cascadia subduction story. Experts at the top of their fields would still doubt the seismic potential of Cascadia for another eight or nine years.

The spectacle of Mount St. Helens was riveting, no doubt about that. It was also a distraction, such a stunning assault on the senses that few, if any, stopped to think about what this eruption might mean in a larger perspective. Rereading Walter Sullivan's story, I then found a clue to why some scientists were able to separate the volcano itself from the much wider threat a magnitude 9 megathrust quake spread across five major cities would pose.

The scientists Sullivan consulted for his pre-eruption feature story had told him how the Cascadia Subduction Zone was thought to be a special case, different somehow from all the other continental collisions around the Pacific Rim. Cascadia's volcanoes do form a line roughly parallel to the coast about a hundred miles (160 km) inland, and the mountain cones are spaced about forty-five miles (70 km) apart, like most other subduction zones with volcanic arcs. But in the minds of skeptics, that's where the similarities ended.

Cascadia is not typical, Sullivan wrote, because "no coastal trench cuts in to the sea floor" at the point where the two tectonic plates converge and "no sloping zone of earthquakes" marks the descent of a seafloor slab beneath the coast. According to Sullivan's sources, the Cascade volcanoes seem to have been created by an east-driving portion of the Pacific floor that had somehow run out of steam. The subduction process, wrote Sullivan, "is no longer vigorous enough to sustain a coastal trench and cause frequent earthquakes."

So Cascadia's smoking gun had run out of ammunition. No deep trench offshore and no deep quakes along the plate boundary—all because the movement of the eastbound Juan de Fuca plate had slowed down or even stopped. At least that's what some experts thought at the time.

Trying to imagine how these huge plates float and slide over the curved surface of an imperfectly spherical planet, geophysicists came up with a sequence of events—a long geologic history—that seemed to fit the observable facts. As the floor of the Pacific Ocean spread apart along the Juan de Fuca Ridge, pushing the Juan de Fuca plate eastward, the rest of the Pacific plate (out on the western side of the ridge) was not moving due west but rotating in a more northerly direction toward Alaska.

At the same time, the North America plate was pushing westward and riding up over top of the eastbound Juan de Fuca slab. Around ten million years ago, so the theory went, the Juan de Fuca Ridge and plate started rotating clockwise, almost as if it were being spun by the angular movement of the two larger plates on either side of it. Think of a car going eastbound through an intersection when a northbound car passes just behind it, clipping the back fender. That northbound motion would make the eastbound car spin to the right, just as it got hit head on by a big westbound truck.

Five million years later, the Explorer plate had broken off the northern end of the Juan de Fuca, and some thought it might have fused or welded itself to the larger continental plate just north of Vancouver Island. There was further speculation that the Olympic Peninsula, on the northwest corner of Washington State, might also have been a piece broken off the Juan de Fuca plate and that it too had been fused to the continent, pressed against the outer edge of Vancouver Island, forcing up the Olympic Mountains in the process.

Two and a half million years later, on this hypothetical timeline, the North America plate had "disposed of" (subducted or recycled) a huge portion of the original Juan de Fuca plate. Down at the southern end, meantime, the smaller Gorda plate was breaking away as well and the spreading ridge offshore—the entire undersea mountain range—had rotated or been spun even farther to the right. At some point, according to this scenario, the relentless westward movement of North America would completely override what was left of the Juan de Fuca plate—and its spreading ridge offshore—just as it had apparently already done farther south in California.

Eventually, with the Juan de Fuca ridge and plate system gone, the boundary between the North America and Pacific plates would become a much simpler structure, an almost straight-line fault starting with the San Andreas in southern California, extending north along the coast, and connecting with the Queen Charlotte fault system all the way to Alaska. Then the dominant tectonic force in the Pacific Northwest would be a more straightforward, northerly compression caused by the northbound drift of the Pacific plate—just the way it is now along the San Andreas in southern and central California. The eastward subduction of the Juan de Fuca and Gorda plates underneath the continent would become ancient history and Cascadia's tectonic threat would be rendered harmless.

In October 1972, Robert Crosson, a seismologist and professor at the University of Washington, wrote a paper suggesting that this had probably already happened. Based on data from the Pacific Northwest Seismograph Network, a newly installed, six-station, high-sensitivity telemetry system capable of pinpointing even the smallest tremors, Crosson and his colleagues had shown that nearly all the jolts in the Puget Sound region around Seattle and Tacoma were the result of north–south compression. There was no evidence of eastward pressure from the Juan de Fuca plate at all, as far as they could tell.

The vast majority of the recent Puget Sound earthquakes had been relatively shallow ruptures in the upper crust of the North American continental plate. The absence of a down-sloping zone of much deeper shocks (known as a Benioff zone) along the eastward-dipping plate boundary and the lack of any recent volcanic activity in the Cascades (in 1972) could be seen as further evidence that the Juan de Fuca plate had stopped moving or was in its final phase of subduction. Without that constant eastward shove from the Juan de Fuca Ridge, the dominant tectonic pressure would have become northerly—and that's indeed what the new seismographic data in Washington State seemed to confirm.

Just across the border in British Columbia, however, two scientists working for the Geological Survey of Canada had looked at their data and come to exactly the opposite conclusion. Robin Riddihough and Roy Hyndman argued in August 1976 that subduction was still happening. They pointed to the "significant eastward dip" of the ocean floor and to the layers of sediment—two and a half miles (4 km) thick, lying on top of the Juan de Fuca plate—that had been dragged sideways into a shallow, less obvious trench at the edge of the continental shelf where they were crumpled, folded, and fractured, all relatively recently.

The continental shelf itself had been recently deformed and uplifted, just like those terraces (former beaches) hoisted up near Cape Mendocino in California. They cited a higher than normal heat flow from the inland Cascade Range that was probably caused by upwelling magma from the melting oceanic slab. All of these were classic symptoms of active subduction, according to Riddihough and Hyndman, although they agreed it would be hard to tell if and when a plate had stopped moving in the recent past. So there remained a degree of uncertainty about whether or not Cascadia posed a clear and present danger.

Another possible explanation for the lack of large earthquakes came in June 1979, when Masataka Ando of the U.S. Geological Survey and Emery Balazs of the National Geodetic Survey suggested that the Juan de Fuca plate was still subducting, but doing it *aseismically*—without earthquakes. Given the lack of large ruptures over the 140 years since white settlers had arrived and written records had been kept, two things were possible. Silence along the boundary zone could either mean the two plates were now locked together by friction and strain was building up for a major rupture, or that big temblors simply didn't happen in this subduction zone. Which brings us back to the idea that Cascadia is somehow a special case.

"In some subduction zones, such large earthquakes do not occur," wrote Ando and Balazs. They had a hunch that friction between the Juan de Fuca and North America plates was too low for the rocks to get stuck together. If, for whatever reason, they don't get stuck—because of a slower than normal rate of motion, perhaps, or a shallow angle of subduction—then movement could keep happening without major quakes. Strain might build up enough to compress and bend rocks in the overlying plate and still not cause a rupture. And they figured the only way to find out for sure would be to measure the rate of deformation along the highways of Washington State.

The first precise leveling survey of Washington's roads had been done back in 1904. By 1974 new surveys had been carried out on ten different sections of highway, some of which ran east–west across the Coast Range mountains. In the time between the first and second surveys, the surveyors' data showed that the outer coast had been lifted upward and the inland areas east of the Coast Mountains had subsided. In other words, the entire mountain range was tilting slightly toward the east and this had to be a result of active, ongoing subduction because it had happened in the past seventy years, not millions of years ago in geologic time.

However, another important detail made Cascadia different, according to Ando, who had recently studied strain accumulation in the Shikoku area along the east coast of Japan. There the Philippine Sea plate is thrusting under the Asian plate—beneath the islands of Japan—along the Nankai Trough. The geologic setting is very similar to the Juan de Fuca Subduction Zone. The dip angle of both subducting plates is a shallow twenty degrees and both oceanic plates are relatively thin. The significant difference is that the outer coastal landmass in Japan is tilting down *toward* the ocean rather than leaning inland as it appears to be doing in Cascadia.

Bending the outer edge of the coast downward as the ocean floor scrapes underneath it is a sure sign the plates are locked together by friction and building strain for a large quake, according to Ando's analysis. Once the rocks along the locked portion reach their breaking point—when friction between them is no longer enough to keep the two plates stuck together—the strain is released in a massive shockwave. As the two plates rip apart in a typical or "normal" subduction zone like the Nankai Trough, the outer coast snaps free from the downgoing oceanic plate and springs back upward. The area slightly inland from the coast subsides at the same time. This is exactly what happened in previous large quakes in Japan, Alaska, and Chile.

In the aftermath of these giant jolts, as the overlying continental plate settled back down to its more or less normal position, some of the coastal uplift remained. In other words, the beach never quite got back to where it used to be because the underthrusting oceanic plate was still down there, still moving below the continent, still causing a certain amount of *residual* deformation. The three-step sequence, according to Ando and Balazs, starts with coastal *down-warping* just before the quake, followed by *heaving upward* during the rupture, and then a certain amount of *residual uplift* of the beach zones in the aftermath.

Cascadia, however, seemed to be doing something entirely different. If the aseismic hypothesis were true, then the Juan de Fuca plate would be just creeping down underneath the continent, slowly and continuously, lifting and tilting the Coast Range mountains to the east, and doing so without getting completely stuck and without accumulating enough strain to cause a major rupture. To me this sounded like the good news. The bad news came in the concluding paragraphs.

Studies of other aseismic zones had revealed that temblors are still possible even if the two plates are not completely locked together. Hiroo Kanamori at Caltech found that if you look at the total distance—how much long-term horizontal movement there had been along the subduction zone in the Kuril Islands, for example—and compared that to the movement that happened during large thrust earthquakes, the ruptures accounted for one-quarter of the total slip. In northern Japan another study showed that quakes accounted for one-tenth of the total subduction rate. Which could mean that even in a mostly aseismic zone—where 75 or even 90 percent of the plate motion is slow, smooth, quake-free creeping—the plates can still get locked together and mega-thrust events do eventually happen. So Cascadia is not completely off the hook for damages, even if the aseismic theory is true.

The tip-off, according to Ando and Balazs, should occur whenever we see the outer coast of the Pacific Northwest start to dip down and get pulled under by the Juan de Fuca plate. Not surprisingly, they recommended constant vigilance by a team of surveyors with state-of-the-art equipment to spot any change along the beach. In the meantime, because the Coast Mountains are now tilting eastward instead of toward the sea, they assured us that a large thrust earthquake from Cascadia's fault is "not expected in the near future."

Then along came the eruption of Mount St. Helens one year later. How could a violent explosion like this not be the sign that convinces all and sundry that Cascadia is still very active and that a tectonic disaster is looming? I suppose the first and simplest explanation was that the debate about Cascadia as "a special case" was happening mostly within the confines of the science community. The general public was not reading these new technical papers, not attending the scientific meetings, and therefore they did not know, for the most part, that Cascadia's fault even existed.

But why did so many scientists who had read the new literature still hesitate?

Gary Carver, who was still mapping thrust faults in the rumpled hills around the Humboldt Bay nuclear plant at the time Mount St. Helens blew, knew that the majority of scientists were skeptical of the Cascadia disaster scenario. Why would the volcanic eruption not have been seen as proof positive of active subduction? He told me that an eruption could still happen even after subduction had stopped.

Presumably, if the Juan de Fuca plate stopped moving tomorrow, the segment of the down-going slab that had already been pulled toward the earth's hot interior would have begun to melt. Plumes of magma would already be rising up beneath the arc of volcanic mountains. So there could be a lag of who knows how many years—hundreds, maybe thousands—between the end of Cascadia's plate motion and the final eruption of Mount St. Helens or one of its neighbors. From that perspective, the St. Helens blast didn't prove anything.

Apparently what everyone needed and wanted was forensic evidence that there had been specific Cascadia earthquakes at specific times in the past. Not just hypothetical scenarios, not just signs that the beaches had been hoisted or the mountains tilted, or even that there had been smaller fractures in the continental crust near the California coast. The only thing that could finally put an end to all the back and forth would be tangible signs of past ruptures along the *entire* subduction zone. And once again, a clue about where that proof might be found was layered in the story written by Walter Sullivan of the *New York Times,* just before Mount St. Helens blew.

The so-called missing trench had to be significant somehow. There were trenches at all the other converging plate boundaries, so why would Cascadia's subduction zone be different? From what I'd read in the science journals, there was a bit of a trench off the west coast, although it was shallow compared to the others and filled with sediment. Why was the down-going angle of the Juan de Fuca plate almost horizontal while the others were steeper and deeper? And what did it matter that the crack was full of mud? The never-ending dump of sand and silt from the turbulent Columbia River and many others along the Pacific Northwest coast had all but buried the fault, so as with the Juan de Fuca Ridge, nobody knew the trench was there at first.

Trenches at other plate boundaries in deeper parts of the Pacific are located far enough away from the outflow of big mountain rivers that silt and sediment can't hide the evidence of subduction. Cascadia's close proximity to the gushing plumes of mountain run-off had created this accretionary wedge, a blanket of muck two and a half miles thick (4 km) that not only filled the crack between the two converging plates but also made the subduction zone nearly impossible to study. There was something else, though—something buried in the sediment—that would reveal the hidden story of Cascadia's past. And it too came from that chain of smoky volcanoes.

Without knowing that Mount St. Helens would soon explode, Walter Sullivan had asked scientists about the eruption of another famous Cascade volcano, just to provide a frame of reference. He wrote about the explosion of Mount Mazama 7,700 years earlier, an apparently world-changing cataclysmic event. Sullivan described it so that readers could visualize what would happen again someday in the Pacific Northwest.

Roughly a hundred miles (160 km) east of the Pacific coast, Mount Mazama, like Mount St. Helens, had been created by Cascadia's oceanic plate subducting underneath North America. This ancient stratovolcano was given its name posthumously, because it had exploded and was long gone before geologists arrived on the scene seven millennia later to piece together what happened. The upper part of the mountain had completely disappeared in a spectacular blast that caused the lower walls of the volcano to collapse inward, creating a huge, circular hole in the ground—a caldera—five miles (8 km) wide. This gradually filled with snowmelt and rainwater to form Crater Lake—with a maximum depth of 1,958 feet (597 m), the deepest lake in the United States.

Magma spilled from cracks along the shattered volcanic rim and surged downhill in avalanches that filled nearby valleys with up to three hundred feet (90 m) of hot rock, pumice, and ash. Somewhere between eleven and fourteen cubic miles (not cubic yards, *cubic miles*, or 46–58 km³) of magma was ejected. A towering column of ash thirty miles (48 km) high rained down for several days on eastern Oregon, Washington,

Idaho, Montana, Nevada, and southwestern Canada. An ash layer half an inch (1 cm) thick was measured in Saskatchewan, 745 miles (1,200 km) from its origin.

Sullivan quoted Grant Heiken, a volcanologist at the Los Alamos Scientific Laboratory in New Mexico, who suggested that "a safe distance from which to watch such an event might be the Earth's orbit of a space station." In more recent times, the U.S. Geological Survey website referred to the Mazama blast as "the largest explosive eruption in the Cascades in the last one million years." Mazama's blast was forty-two times more powerful than Mount St. Helens' in 1980. Only the explosion of Krakatoa (recounted by Simon Winchester in his excellent book of the same name) off the coast of Indonesia in 1883 could compare to Mazama's magnitude and impact.

Eventually Mazama's ash became famous in its own right. Geologists used radiocarbon dating to determine when the eruption had occurred and then, wherever researchers spotted the recognizable Mazama layer, it became an important stratigraphic marker, a distinctly visible line that appeared in bands of sediment all over western North America. Mazama provided geologists and archaeologists with a key reference point for geologic calendars and timelines. If you could find Mazama ash in a drill core sample, you could tell roughly how old the layers above and below it were and in what order major geologic events had occurred.

And that's how Mazama became a key factor in the next phase of the investigation of Cascadia's fault. Within eighteen months of the eruption, much of the ash had been carried away by rain, washed downstream by hundreds of creeks and dozens of rivers, and dumped in great heaps along the offshore continental shelf and in deep-sea canyons that wandered for hundreds of miles along the coast. Thousands of years later, a team of marine geologists from Oregon State University would find the distinctive Mazama line in cores of mud they gouged from the ocean bottom—cores that may have looked insignificant at first glance. On closer examination, it turned out they contained the elusive, muddy fingerprints of Cascadia's violent past.

PART 2 SETBACKS AND BREAKTHROUGHS

CHAPTER 9
Mud Cores and Lasers: The Search for Evidence

At five o'clock in the morning it was still pitch black out on deck and the heavy air promised another sticky day. The boatswain, a muscular young guy in his mid-thirties with a patchy beard and a sunburned face, instinctively stepped back to get clear of the bight of steel cable at his feet.

He glanced up behind the blue-green glare of the floodlights to make eye contact with the winch operator in a glass booth on the upper deck, then raised his right arm and made a circling motion, fingers pointed upward at the sky. The diesel roared a little louder, the cable drum began to turn, and a heavy metal shaft rose from the shadows.

A handful of scientists gathered along the starboard rail for a final inspection of their strikingly low-tech research probe—a "Benthos gravity corer." To the unfamiliar observer, it looked like nothing more than a hollow, vertical steel tube, about ten feet long and four inches in diameter (3 m by 10 cm), fitted at the top with a collar of five lead weights shaped like doughnuts and capped off with a set of angular fins, the kind you used to see on bombs, designed to make the rig fall through the water straight and true. Gravity would soon take it to the briny deeps, where it would stab the sea floor and try to capture nearly ten thousand years of tectonic history.

The scientists, a mixed team of marine geologists, oceanographers, graduate students, and veteran researchers from the United States, Britain, Spain, Belgium, Germany, Japan, and Indonesia, had crossed the Andaman Sea aboard the *Roger Revelle,* a research ship operated by the Scripps Institution of Oceanography. They had sailed from Phuket, Thailand, on Monday afternoon, May 7, 2007, slipping between the Nicobar Islands and the northern tip of Sumatra into the Indian Ocean. Just before sunrise on Wednesday they were about to drop their first probe into the ocean mud west of Banda Aceh, that unfortunate beachfront town so memorably wrecked by the earthquake and tsunami of December 26, 2004.

I had met U.S. chief scientist Chris Goldfinger, a marine geologist from Oregon State University (OSU), a year earlier in 2006 while filming an update to my original 1985 Cascadia documentary. He and his team had been punching core samples out of the sea floor and then adapting a technique used by oilfield geologists to match up the stratified layers of clay, silt, and sand at various points along the rupture zone for clues about the size and timing of past earthquakes. In the spring of 2007, he and an international team of scientists went to sea off Indonesia to find out what the latest event—the magnitude 9.2 Sumatra–Andaman megathrust—might tell them about seismic patterns in Cascadia.

In the ship's lounge the previous afternoon, he had gathered the graduate students, postdoctoral researchers, and coring technicians for an introductory briefing. Looking more like a veteran California surfer than a labcoated investigator, Goldfinger clicked his mouse and the first image appeared on a roll-up screen: a 3D cutaway view of the sea floor off the Oregon coast. "So here's what Cascadia looks like in cross-section," said Goldfinger. "Just a very simplified image of the accretionary wedge and the forearc structure."

I had seen this seafloor display at his office in Corvallis several months prior to the voyage and thought of it as a magical mystery tour of a place I'll never get to visit in person. Cascadia's undersea topography (bathymetry is the correct term) was a hidden landscape, another whole world, exotic yet oddly familiar. The same rugged terrain you see above ground along the Oregon coast exists in a parallel universe on the ocean floor.

This deep-sea world had its own substantial hills and valleys, its own cliffs and steep canyons, and what looked like sharply cut river channels running across a wide, flat prairie that vanished in the dim, purple distance at the back of the 3D image. When Goldfinger was at his main computer terminal on campus, he could fly around inside the software, touring the canyons, floating over the folded hills of piled-up sediment (the accretionary wedge) in a trip along the subduction zone that was better even than my childhood memory of Captain Nemo's fantastic voyage.

"The main point here," Goldfinger continued, "is that the rupture of a great earthquake in a subduction zone is almost always completely underwater." He pointed to a fine line at the base of a row of hills along the sandy bottom roughly sixty-five miles (105 km) offshore, where the Juan de Fuca plate slides beneath the edge of the continent. "For other faults, like the San Andreas, that are exposed at the surface, you can walk right up to it," again he paused, "get out your rock hammer, and tink, tink, tink. Or get a backhoe and dig a hole across it." His punchline was that studying Cascadia is a tad more difficult because there is no way to get a backhoe or even a hammer to the bottom of the continental slope beneath thousands of feet of seawater.

Goldfinger changed slides. "So what we're gonna do on this cruise," he explained, "is turbidite paleoseismology—based on the idea that if you have a big enough earthquake, you're gonna generate a lot of land-slides on the submarine margins. And you basically can just go to the bottom of the hill, take piston cores," he clicked the mouse again, "and you should get a vertical record of landslides." The new image onscreen showed a cutaway view of a core sample sliced open on a laboratory workbench. A vertical slice through the stratified layers of thousands of years of seafloor mud, sand, and silt.

One of the younger researchers raised his hand. "Define turbidite."

Goldfinger grinned. "A turbidite is a sandy, muddy, high-energy deposit coming from submarine landslides." He clicked the mouse again to a closer view of the core sample. "It's an underwater sediment plume, gravity driven." He waved a laser pointer at the screen. "Because all this material is entrained in the flow, it's subject to gravity and it's gonna keep going downhill until it runs out of—downhillness." He wiggled the laser dot at a series of horizontal lines in the core, each one representing a different landslide.

"The trick is," said Goldfinger, "how do you determine if the landslides are earthquakes?" Meaning the sediment might pile up for years at the head of an undersea canyon and then simply collapse under its own, unstable weight. Or it could be knocked loose in the relatively shallow water of the continental shelf by turbulence from a big storm passing overhead. But Goldfinger and the team from Oregon State were pretty sure those dark lines in the ocean cores were the physical remnants of Cascadia's tectonic past, debris from landslides that had been triggered by big seismic shocks.

Out on deck, the boatswain signaled to the winch operator, who swung the boom out over the starboard rail where the steel piston hung in a stiff breeze. He then leaned over the side to confirm that the acoustic "pinger" was powered up, a smaller metal tube now clamped to the cable just above the piston probe. It would send signals to the ship's multi-beam sonar system as the coring rig approached the ocean floor. Satisfied that everything was good to go, the boatswain stepped back again, looked over his shoulder at the crane operator and raised his right arm, thumb down. At the working end of the boom, a pulley block began to spin and cable rumbled off the drum. Moments later the piston and its pinger disappeared with a quiet swish into the choppy black water.

Two hours later, in the main laboratory control room, all eyes focused on flat-panel screens and digital readouts as the piston rig dropped through the dark abyss near the bottom of the Sunda Trench. Numbers clicked over rapidly. More than two and a half miles (4,100 m) of cable had spun through the block when Chris Moser, one of the senior coring technicians from OSU, punched the intercom. "Okay, Eddie, stop the winch."

After waiting a few moments for the rig to stabilize, he gave the order to spool it down the rest of the way. The piston hit bottom at 14,400 feet (4,380 m). It punched a five-foot (1.5 m) hole in the mud at the mouth of the seafloor canyon closest to the Aceh rupture zone— where the earth had begun to rip apart in the monster quake of 2004.

It took another two hours to haul the rig back to the surface, so naturally, it arrived after a hazy, overcast sunrise, just in time for breakfast. One of the coring technicians grabbed an industrial-size pipe wrench to unscrew the nose cone from the shaft before the boatswain and a single deckhand slid out a plastic pipe concealed inside the slightly larger metal piston, like an arm inside a sleeve. The see-through plastic tube containing the core sample looked surprisingly light as the two men hoisted it onto a set of stanchions bolted to the deck. Light, because it was almost empty. After more than four hours from deployment to recovery, the ten-foot (3 m) section of pipe contained maybe eight inches (20 cm) of clay and silt—a disappointment, to say the least. They would have to start over.

Later the next day five members of the science team lifted another, larger piston core from the sea. The boatswain and the crane operator hoisted the steel pipe casing away, exposing a twenty-foot (6 m) core tube inside (this one looked like ordinary, white PVC drainpipe). This time they'd captured a good sample. In the main laboratory, they mounted the plastic tube in a set of clamps on a long workbench and ripped it in half lengthwise with an electric saw. Everyone in the lab was buzzing with energy, measuring, labeling, and entering data into logbooks and computers.

Chris Goldfinger watched as Russ Wynn from the National Oceanography Centre in Southampton, England, dragged a putty knife across the surface of the new sample, skimming off the top layer of runny mush to create a smooth, flat finish. Now it was easier to see the four or five horizontal stripes of darker, lumpier material that stood out against the bland, gray goo of deep-sea mud. These brownish smudges presumably were the killer's fingerprints—the turbidite layers from undersea landslides triggered by the deadly Sumatra temblor of 2004 and perhaps from a subsequent rupture in 2005.

Chatting later with a few of the grad students in the ship's lounge, Goldfinger began telling the tale of how this kind of research—this deep-ocean version of earthquake hunting, cross-bred with oilfield exploration techniques—got started at OSU. He described how his thesis advisor, geology professor LaVerne Kulm, "took cores in the late '60s along the Cascadia margin. And remember, in 1968 plate tectonics was only four years old at the time. So people were just kinda getting used to the whole idea."

"They were taking cores out here," said Goldfinger, pointing to a place on the map more than sixty miles (100 km) to sea, southwest of the Columbia River estuary, "and they noticed that there was an ash deposit out there called the Mazama ash." This gave Kulm and his team a timeline, a starting point from which to gauge the age of the other deposits, those darkish lines of turbidite debris they had found in the vertical column of marine mud.

Thirteen clearly defined turbidite deposits had been found in core samples taken many miles apart along various riverlike canyons on the ocean bottom. "Here, here, here, and here," pointed Goldfinger, all over the map. "All of these places had thirteen turbidites. So one afternoon, as Vern tells it over beers, he and his students were puzzling about *why*—why would all these cores have the same number of turbidites? These come from different river systems, different parts of the margin, different geology. They have absolutely nothing in common except, apparently, they had the same number of turbidites deposited in these cores."

Goldfinger was warming to his subject. "And the way Vern tells it, one of the students goes: 'Hey, maybe it could be earthquakes! And then they would just be triggered all along the whole margin. And you'd get the same answer wherever you went.' And Vern said, 'Nah, nobody'd ever believe it.'" Goldfinger looked around the room, paused for effect, and then added, "Well, turns out they were right."

At the very least LaVerne Kulm was right to suspect that in 1970, when several scientific papers based on those first Cascadia mud cores were published, nobody would believe thirteen giant earthquakes had caused thirteen landslides on every major offshore river channel along the Oregon coast. Chris Goldfinger didn't believe it himself. At least not at first.

Scientists generally don't trust coincidence, and the absence of a convincing explanation guaranteed that the story told by those mud cores was not over. The final chapter—the way those mysterious layers of gunk from the sea floor eventually changed the history of Cascadia's fault—started to make more sense to me only after I'd dug up a little background on where all that silt and debris had come from in the first place.

Standing on the South Jetty at the mouth of the Columbia River on a stormy day in January, a person would have to squint to make out details along the far shore. The Cape Disappointment lighthouse would flash a smeary streak of white through sheets of gray, horizontal rain. Then, with another gust, the winter gloom would close in again, making it difficult to see the state of Washington from the Oregon side. At this point, the river is four miles (6.4 km) wide.

A few hardy souls, storm watchers who love nothing more than the wet sting of salty wind on their cheeks, can nearly always be found here clinging to the guardrail around the viewing platform at Fort Stevens State Park, their exhilarated or foolish companions leaning into the gusts and grinning at the ferocity of it all. The Columbia River's freshet fighting an incoming Pacific tide is an impressive sight.

Unlike most other big rivers, the Columbia has no delta to speak of, no meandering Mississippi mud flats, no maze of trackless swamps. Hemmed in by hard rock mountains that were pushed up by plate tectonics, the river chews like a chainsaw through steep, stony canyons and then gushes headlong into the open Pacific. Columbia's estuary is more compact than a river this size would normally be. Much of the mud that might have formed a sprawling delta in less vertical terrain gets dumped almost directly into the sea.

Suspended in its swollen belly, the river carries silt, sand, and forest debris from 258,000 square miles (668,000 km²) of western mountains that are slowly dissolving under millions of years of rain, rock-splitting ice, and heavy snow. Along the way a host of smaller streams and rivers—the Kootenay, the Pend Oreille, the Snake, the Kicking Horse, the Kettle, the Cowlitz, and many others—add their contributions to the flow and the silt load. At Portland the Columbia inhales the Willamette River, swings north to find a gap between the Coast Range and Olympic Mountains, and then turns west one last time for its final plunge to the sea. The big river's muddy torrent sweeps past the outport town of Astoria, dumping about 265,000 cubic feet (7,500 m³) of fresh water into the snarly North Pacific every second.

The ocean fights back with tremendous swells, pounding surf, and a tidal surge that will not be denied. At the point roughly five miles (8 km) offshore where these two irresistible forces meet and do battle—the infamous Columbia Bar—a never-ending churn between outgoing river and incoming tide, amplified by howling winds—creates standing waves as high as thirty feet (9 m). Having reached the horizontal plane of the sea after more than a thousand downhill miles (1,600 km), the river slows and begins dumping its cargo of suspended debris in a constantly shifting barricade of sandbars and shoals that make the mouth of the river one of the most hazardous places in the world for mariners. Some say nearly two thousand ships and boats have been wrecked along this stretch of the coast—nicknamed the Graveyard of the Pacific—over the past two centuries.

Once the Columbia has forced its way over the bar and out to sea, it veers north, pushed by the Davidson Current and huge Pacific storms. Along the southwestern shore of Washington State, the river dumps more of its load, giving birth to long spits of sand and narrow barrier islands parallel to the mainland with miles and miles of beautifully isolated beaches.

In satellite photos you can see a muddy plume swirling across the edge of the continental shelf before apparently disappearing into the ocean depths. Losing momentum, the Columbia finally drops what's left of that pulverized mountain debris onto the sea floor, piling it up precariously at the head of a steep canyon. The finer grains of silt keep moving farther out to sea.

Reading two 1970 papers by LaVerne Kulm and Gary Griggs (an OSU graduate student when this story began), I learned why those canyons and channels cutting across the flat surface of the ocean floor looked so much like a big river and its tributaries meandering across the prairies. It's because that's exactly what they are. This spidery web of channels is basically an extension of the Columbia River system across the sea floor—through the mud on top of the Juan de Fuca plate.

When massive glaciers filled the mountain valleys and covered most of the Pacific Northwest more than ten thousand years ago, the level of the sea was roughly 425 feet (130 m) lower than it is now. With so much water trapped in glacial ice on land, the ocean shrank and the continental shelf was exposed as dry land cliffs. In those days the Columbia had to flow much farther west to reach the sea. It ran across the drying continental shelf, cutting a groove all the way to the outer edge. Then it sliced a canyon down the steep slope to the flat oceanic plate below.

As ice sheets came and went over millions of years and the level of the sea rose and fell, so the Columbia periodically changed course. When ice was thick in the mountains and the sea level was low, the river's main current ran directly west across the exposed shelf and down to the sea through what is now known as Astoria Canyon. When the glaciers started melting again, gradually raising sea level and redrowning the canyon, the river's current swung northwest, pushed by strong ocean currents and storms, and meandered across the reflooded continental shelf until it reached the edge again farther north, where it cut another groove down the continental slope at what is now called Willapa Canyon, off the coast of Washington.

During several of these big melting cycles, broken slabs of ice plugged a narrow valley in the upper Columbia basin near the Idaho–Montana border, creating an ice dam 2,500 feet (760 m) high that backed up an inland sea called Glacial Lake Missoula, as big as Lake Erie and Lake Ontario combined. Eventually the dam broke—twelve thousand years ago—releasing a catastrophic torrent of water and debris that roared down the channel cut by the river, spilling out to sea across the continental shelf, then gushing down Astoria and Willapa Canyons before spreading like thick mud soup across the Juan de Fuca plate.

That's why the Cascadia basin—the spreading section of ocean floor otherwise known as the Juan de Fuca plate—came to look so much like an underwater prairie. It got buried repeatedly by glacial outwash and periodic floods until the ocean floor was 1.9 to 2.5 miles (3–4 km) deep in silt, sand, and debris.

When I read about this, I recalled the "no trench" part of Walter Sullivan's story in the *New York Times*. Evidently the subduction zone was so full of Columbia sediment and Missoula mud that the first wave of geologists and oceanographers to explore the Cascadia basin with echo sounders could not see much of a trench where they thought one should have been. The early technology used for mapping the sea floor could not penetrate the mud to see the real basement: the boundary zone where the Juan de Fuca plate was dipping down and thrusting its way underneath North America.

With no trench obvious, Cascadia became "a special case," unlike most other subduction zones. Add to this the absence of big temblors and it must have seemed like all the proof any reasonable person might need to be convinced Cascadia's fault was aseismic—quake free and essentially harmless. But when Griggs and Kulm and the team at OSU began to investigate the web of riverlike channels crossing the deep ocean floor, the hypothesis of aseismic subduction started to unravel.

They used heavy piston rigs to gouge core samples that showed a long series of landslides of silt and sand (those turbidity currents that Goldfinger explained years later aboard the *Roger Revelle*) that had flowed down the canyons and been deposited on top of the older Missoula mud. One of those layers was the infamous Mazama ash, which allowed them to radiocarbon date all the other turbidites and calculate the average amount of time between landslides—which turned out to be roughly 550 years. The timing between slides seemed unusually consistent. Another coincidence? Not likely, yet nobody knew how to explain what appeared to be a recurring cycle. Thirteen turbidite landslides—roughly 500 to 600 years apart. Why?

The core samples also painted a vivid picture of what happened once these debris flows began to move downhill. Griggs and Kulm had calculated that each year about one million cubic feet (28,000 m³) of muck was being carried across the continental shelf by the Columbia and dumped at the head of Willapa Canyon. So a million cubic feet of this stuff piles up every year for five centuries—and then *something* makes it tumble.

The core samples traced the downhill flow of these currents and showed that some were so high and fast they had splashed over the walls of the main deep-sea channel and spread out sideways as much as 10 miles (17 km). The biggest flows were more than 325 feet (100 m) high and ran more than 400 miles (650 km) down the channel. At approximately 30 feet or more (10 m) per second, one of these swirling plumes would take nearly two full days to run its course—to run out of "downhillness," as Chris Goldfinger had put it. All of which was amazing enough, although the key question—what had triggered the landslides so regularly for thousands of years—remained unanswered.

Griggs and Kulm offered two possible causes, "periodic earthquakes or severe storms." They drew no conclusion of their own. As Goldfinger told their story years later to a newer generation of graduate students off the coast of Sumatra, nobody in 1970 would have believed the earthquake hypothesis because there had never been a big subduction shake in the Pacific Northwest in all of recorded history.

The logic sounded straightforward: if these kinds of megathrust events were possible, we would have seen one by now. Surely in 150 years of recorded history one of these monsters would have attacked. There was a well-accepted principle in geology called uniformitarianism, which held that "the present is the key to the past." Geologic processes that we see happening now are the same processes that happened long ago. Therefore, if we see no great earthquakes in Cascadia now, this subduction zone has probably always been quiet.

Without more data, it was simply easier to believe that some howling great winter storms had triggered all those offshore landslides. But every 550 years? How could anything in nature be so apparently punctual? That part still rankled for those who were suspicious of coincidence. And there was one doubter in particular who just wouldn't let it go.

Seismologist John Adams, whom I'd met at the Pacific Geoscience Centre on Vancouver Island in 1985 while filming my first earthquake documentary, already knew that plate boundaries could take several centuries to build up enough strain to rupture. Before moving to Cornell University in New York in the late 1970s to work as a postdoctoral research associate, he had completed a study of the Alpine fault, along the southwest coast in his home country of New Zealand. There, instead of subducting, or diving underneath, the Pacific plate was *obducting*—being forced over the top of the Indo-Australian plate.

Like Cascadia, there had been no major ruptures of the Alpine fault system in all of recorded history, which in this case amounted to roughly 150 years. There were plenty of signs, however, that land along the mountainfront had been folded and bent and was under extreme stress. Beaches on the Pacific plate had been pushed up into terraces as they had in Cascadia. The Southern Alps mountain chain on New Zealand's southern islands had been uplifted as a result of compression between the plates, and in places Adams was able to study both vertical and horizontal displacements caused by earthquakes that happened centuries ago.

He noticed several other important things about the timing and the amount of movement along the Alpine system. Some—not all—of the built-up strain had been relieved during big ruptures that happened in the not-too-distant past, but there appeared to be "seismic gaps." It was pretty obvious that parts of the fault were moving spasmodically in earthquakes. Other segments of the fault, however, showed no evidence of rupture and were either sliding along smoothly or had been stuck together by friction, building up stress for a long time, and they were probably ready to slip again in a big quake. Seismologists call this a "stick–slip" scenario.

Just as he would find years later in Cascadia, Adams learned that the science community was divided about the risk posed by the Alpine fault. Even though it was a "San Andreas–scale" crack in the crust, few seismologists had paid it much attention. "They didn't see earthquakes," said Adams. "Their seismic hazard analysis actually ignored it, basically. Whereas the geologists said, 'This thing has moved recently. You can see the offsets and the other characteristics. And therefore, it has to be an active plate boundary and will generate great earthquakes.'"

The long gaps between rock- and landslides triggered by Alpine earthquakes were eerily similar to the intervals between the deep-sea landslides that Griggs and Kulm had found in the mud off the Oregon coast, so it's easy to see why Adams was intrigued by their papers when he finally came across them. While Griggs and Kulm weren't really looking for quakes (they had set out to study the structure and evolution of the deep-sea channel system), serendipity gave them data that would later play a pivotal role in the debate about Cascadia's fault.

Adams, on the other hand, was definitely searching for seismic fingerprints—earthquake history—which is why he would eventually write to Oregon State University asking permission to examine the mud cores, data logs, and timelines compiled by Griggs and Kulm, Hans Nelson (who would later team up with Chris Goldfinger on a series of follow-up studies), and other members of the original OSU team. Adams wanted a closer look at the patterns. In the meantime, he drew a chilling conclusion about the Alpine fault zone in New Zealand.

While there was ample evidence of seismic activity to the north and south, there had never in recorded history been a major rupture along the central part of the fault zone. Now, with physical evidence from a series of dated landslides, Adams felt confident more of the same would occur. He wrote that large quakes with "a rupture length of 270 kilometres [168 miles], a maximum displacement of 9 metres [30 feet], and magnitudes of approximately 8 are indicated for the central part of the Alpine fault." New Zealand, like Cascadia, was apparently locked and loaded for a major shockwave.

By the time his Alpine paper was in its final draft and being peer reviewed in the winter of 1979, Adams was already working in North America with a keen interest in the Cascadia Subduction Zone. The Alpine paper was published in a scientific journal called *Geology* only two months before the eruption of Mount St. Helens, and I suspect its significance—especially the parallels to Cascadia—may have been lost in that spectacular volcanic dust cloud.

Unfazed, Adams continued to work on other evidence that Cascadia was an active threat. He had ongoing battles to fight against conventional wisdom and the principle of uniformitarianism. While at Cornell he had begun working with a researcher named Robert Reilinger on a study of how much the Coast Range mountains of Washington and Oregon were tilting to the east. Like the work of Ando and Balazs, which had come out in 1979, this Cornell project involved new data from highway survey crews that showed a significant upheaval: a change of elevation along the entire western side of the mountain range.

In 1982 Reilinger and Adams took the Washington highway data from the earlier study and extended it southward by adding new measurements from five more resurveyed east–west highways crossing through the mountains in Oregon. They showed that survey markers located near the coast had been lifted up a noticeable amount in the less than eighty years since the last set of surveys. Tide gauge data along the coast showed pretty much the same thing; the beaches had been lifted as well. Put it all together and you got a picture of a mountain range about 370 miles long and 37 miles wide (600 km by 60 km) being hoisted up along its western edge and tilted, en masse, toward the east.

In this new mountain-tilting paper Adams and Reilinger drew attention to another worrisome study published only six months earlier by Jim Savage and a team at the USGS that showed land in the Puget Sound lowlands around Seattle apparently being compressed: squished together in a northeasterly direction. That was the same direction the Juan de Fuca plate was supposed to be moving. If the ocean floor was actively sliding underneath the continent and *if* the two plates were locked together by friction, this kind of compression, or "crustal shortening," near Seattle was exactly what you'd expect to find. It was also exactly contrary to what Robert Crosson and Ando and Balazs had said earlier.

When Ando looked at the vertical shift of outer coast survey markers and the eastward tilting of the Coast Range, he and Balazs reasoned that the long-term, apparently quake-free uplift meant the two tectonic plates were *not* locked. That's why Cascadia was seismically quiet. So how could one explain Savage's new compression data? How does the ground squeeze together if the plates are not locked?

A series of measurements of the distances between geodetic survey monuments on opposite sides of Puget Sound, spaced six to eighteen miles (10–30 km) apart across the sound, revealed a surprising and somewhat baffling trend. Between 1972 and 1979, Savage and his colleagues used a Geodolite, a powerful and precise distance-measuring instrument that fired a laser beam from a survey marker on one side of Puget Sound across to a similar marker on the other side. There, a bank of highly polished mirrors bounced the laser back to the Geodolite, which measured how long the beam took to make the round trip.

If it took less time in 1979 than it did in 1972 the two markers had to be closer together, and that's exactly what they found. They figured the accuracy of the Geodolite was within 0.2 inches (5 mm) and that the amount of squeezing of the valley floor was statistically significant. Savage and his coauthors (Mike Lisowski and Bill Prescott) recognized that their new data were "not easily reconciled" with Ando's aseismic subduction concept, but they published them anyway. They concluded that the laser measurements were evidence of strain building up, probably caused by the subduction of the Juan de Fuca plate underneath North America.

The significant point was that the two plates *had* to be locked together for this kind of strain to accumulate. Savage and his team were saying as politely as they could that Ando and Balazs must be wrong. "The implication is clearly that the Washington and Vancouver Island coasts are subject to great, shallow, thrust earthquakes," they wrote in June 1981.

A year later, when John Adams was drafting his conclusions for the paper on Coast Range tilting based on highway survey data, he spoke to Savage on the phone and heard his idea about why the mountains might be tilting to the east. In Japan and several other places, subducting plates were bending the outer coast *downward*. Why would the Pacific Northwest coast be different? Why would it tilt eastward? Savage had a hunch that if the two tectonic plates were locked at a point only slightly inland from the coast—if the point of impact was nearly head on and close to the beach instead of deeper down and farther inland, beneath North America—then maybe the Juan de Fuca plate would push the mountains and beaches upward (like a crumpled fender) instead of dragging the coast down and curling it under, like in Japan.

Adams latched on to this as a possible explanation for the eastward tilting and thought he'd made a pretty convincing case. He was not completely successful, however, in exorcising the demon of aseismic subduction. At the request of his more senior coauthor, Robert Reilinger, he wrote a concluding paragraph that equivocated enough to dull the edge considerably. Taken as a whole, the tilting data and the lack of large quakes "suggests that subduction is occurring aseismically, although alternative interpretations are possible," they wrote cautiously. Thinking about it more than thirty years later, Adams had to laugh. "That phrase was largely put in to—shall we say—to take the controversy out of the paper, to make sure it got through the peer-review process."

But Adams was already hard at work on another study that would pull fewer punches. He was like a dog with a bone. He knew the mountains were tilting, he knew Puget Sound was getting squeezed, and he intended to follow up on those beaches that had been shoved up into marine terraces along the Oregon coast. He knew from his New Zealand studies that it took hundreds of years to build up enough strain for a giant subduction quake. And he knew about those turbidite mud cores from offshore landslides that also were roughly five hundred years apart. Were the deep-sea landslides physical proof of Cascadia's violent past? He was determined to find out.

CHAPTER 10
The Whoops Factor: Cascadia's True Nature Revealed

While John Adams watched the Coast Range tilt and Jim Savage tracked the squeezing together of mountain peaks in Puget Sound, Mike Schmidt was learning about a new technology that could make the measurements far more precise. He would eventually join a team of researchers on Vancouver Island, where the distance between several mountain peaks was being resurveyed to find out whether continental drift was shoving them closer together as well.

My first impression of Schmidt was that he'd rather climb a big rock pile than stand there and look at it. He's a bearded bear of a guy who seems to have chosen the right career. In 1992 he led a team of Canadian scientists to the top of Mount Logan, Canada's highest peak and its fastest rising mountain. Fast in geologic terms, it grows by several fractions of an inch each year.

As a mountain climber, geophysicist, and surveying engineer, Schmidt wanted to establish new geodetic markers near the summit and try out some brand new and allegedly portable GPS technology, which was still in the experimental stage at the time, to trace the peak's constant movement. Logan, which occupies a big chunk of the southwestern corner of the Yukon, is poking up and creeping horizontally for the same reason that mountains in Puget Sound near Seattle are getting squeezed together. In the case of Logan, the floor of the Pacific Ocean is jamming itself underneath North America from the Gulf of Alaska.

This is the same tectonic force that caused the 1964 Alaska earthquake. The Pacific plate is thrusting the entire St. Elias Range a tiny bit higher and shoving it slowly inland at the same time. Because 1992 was Canada's 125th birthday and the 150th anniversary of the Geological Survey, Schmidt came up with the idea of putting together an expedition to climb the mountain and settle a long-standing debate about how high Mount Logan really was. With sponsorship from the Royal Canadian Geographical Society, he and his team did exactly that.

They couldn't simply fly to the top because the summit was too high and the air too thin for ordinary helicopters. Plus the researchers themselves would need time to acclimatize to the altitude before starting the hardest part of their work. So instead of an easy ride in a chopper, they made nine trips over three days in a single-engine, ski-equipped Helio Courier airplane to airlift the fifteen members and all their gear to a base camp on the Quintino Sella Glacier, 9,055 feet (2,760 m) up the mountain. From there they had only another 10,495 feet (3,199 m) to go, Schmidt told me, slogging steeply uphill the hard way. It was the only "relatively safe" option.

For thirty days starting in early May they skied and climbed and packed loads of food, tents, sleeping bags, clothing, climbing gear, and heavy cases of the new high-tech survey equipment—satellite receivers, antennas, and heavy batteries—steadily upward through spectacular spring sunshine and howling late-winter snowstorms that nearly forced them back. Being so close to the Gulf of Alaska meant that nasty weather could blast across the slopes with almost no warning. And it did, several times.

By Schmidt's account, though, the expedition was a complete success. They nailed a new brass survey marker at 18,044 feet (5,500 m), on the edge of the Logan plateau; when they finally reached the summit, the portable GPS system worked perfectly and the official height of the mountain was confirmed at 19,550 feet (5,959 m). But they also proved under extremely harsh field conditions that this new, extremely precise technology could be used to help figure out what was really happening along Cascadia's fault.

Measuring mountains and the drift of continents using satellites, sophisticated antennas, and software to track the warping and bending and horizontal migration of land caused by plate tectonics would become the focus of Mike Schmidt's working life. At the Pacific Geoscience Centre on southern Vancouver Island, he helped develop the technology and methodology for tracking the minute and ongoing deformation of the earth's crust. But before GPS, there were lasers and Geodolites and each step along the way was a huge improvement.

I heard the story of how it all began from Schmidt's senior colleague, Herb Dragert, who was there at PGC when the study of migrating mountains began in 1976. Apparently a burning desire to prove that Tuzo Wilson at the University of Toronto was right all along about plate tectonics had been Dragert's personal motivation. As an eager young student, his imagination had been fired by Wilson and those big ideas about drifting continents. "He kind of said to us, 'Okay—we do get plate convergence on the west coast and we should be able to measure the actual motions of the earth's surface,'" Dragert recalled with gusto. "These mountains should be squeezed!" Which is precisely what he, Schmidt, and a team of others from the Geological Survey would try to confirm.

On Vancouver Island the idea was to locate the original stone cairns and brass markers on mountain tops that had been surveyed back in the late 1930s to remeasure the distances and the angles between the peaks and thus find out whether—or indeed, how much—they had moved by the late '70s. Decades later I wanted to see how the work had been done. In 2007 I needed a way to illustrate the process for a television documentary, so Schmidt was going to show me by doing it again.

Thus I found myself in a helicopter once more, flying this time toward the lumpy shoulder of Mount Landalt, a few miles north of Lake Cowichan on southern Vancouver Island. The skids of the JetRanger touched down gently on a bed of gray lichens and green moss sprinkled with tiny, bright fuchsia-colored flowers. Schmidt pulled out the first of two steel cases, each a little larger than your standard, full-size suitcase. Then he carried a set of heavy-duty tripod legs to the summit and set them up directly above the control point, the brass marker that had been established by the original survey crew back in 1937.

The instrument he hauled out of its foam-padded metal case was about twice the size of that old breadbox your grandmother used to have. It was a Rangemaster III, with a bright orange housing, knurled brass knobs, a black instrument panel, and a self-contained digital computer that flashed the distance calculation to an LED readout. State of the art in 1976, it still appeared in perfect working order after many years in a storage locker. The helicopter pilot then flew across to Mount Whymper, the next nearest mountain in the hazy distance, where he landed and set up a reflector box on a tripod directly above the survey marker on that peak.

Back on Mount Landalt, Schmidt peered through his viewfinder on the Rangemaster and used a portable radio to call the pilot, who then tilted the bank of mirrors on Mount Whymper until Schmidt got a return signal—a reflection of the laser beam—on Mount Landalt. Scintillating shafts of cherry-red light bounced off a dozen mirrored prisms in the reflector box. The Rangemaster then performed its magic: a quick calculation of the time it had taken the beam to shoot across the valley from one peak to the other and bounce back. With laser gear like this they could measure the distance between peaks up to twenty-five miles (40 km) apart and be accurate within fractions of an inch. A significant improvement, but there was still another problem to solve.

With nothing more sophisticated than pack horses, climbing gear, and transits (telescopic instruments mounted on tripods for measuring precise, horizontal angles between objects that are far away) the British Columbia survey crew back in 1937 could accurately plot the geometry between a series of peaks. But computing the exact distances between mountains was extremely difficult. In those days most surveyors still used sixty-six-foot chains (eighty chains to the mile, or about fifty to the kilometer) to establish a baseline measurement. If you know the angles and the length of one side of a triangle (the baseline), you can calculate the lengths of the other two sides. But because of jagged mountain terrain and dense bush, the distances calculated and printed on the old maps of Vancouver Island were too imprecise to work as reference points in a modern-day study of minute tectonic creep along a fault. The new laser equipment that became available in the 1970s changed everything.

Dragert and Schmidt and their team used the laser Rangemaster to redraw the original triangles between the peaks and see whether *they* had changed. With the Rangemaster's new measurements they knew exactly how long each leg of the triangle was and could then calculate the precise angles between the peaks. Then they compared the new triangles to the old ones from 1937. When they did—bingo!—they saw that the angles had changed, which meant that at least some of the peaks had moved since 1937. "We proved that the margin was deforming," said Dragert. "The mountains were indeed being squeezed landward," ever closer to the continental mainland.

He pointed to a specific example on the old map, a triangle of dark lines drawn by the original surveyors. The new laser triangle clearly did not match the old one. It was bent out of shape because the mountain closest to the west coast—Mount Grey, a 4,570-foot (1,390 m) peak about halfway up the Alberni Inlet—had been shoved eastward nearly eight inches (20 cm) in less than forty years. Not a huge amount, by the sound of it, but imagine the entire island coastline, hundreds of miles' worth of mountain rock, being pushed horizontally like that. When I thought of another few inches of horizontal movement each year—for 310 years since the last Cascadia earthquake—the total amount of accumulated strain built up along the fault was mind-boggling.

"And that was totally consistent with our expectation," said Dragert, "that if the subduction zone is *locked,* we *have* to see deformation." He poked his finger emphatically at the map again. "And indeed we *saw* the deformation which was the final nail saying, 'Look—this is *not* slipping smoothly. This subduction zone is *locked!*'" One might think that should have been the end of the aseismic subduction idea. But it was not.

For those who knew what the data were saying, the late 1970s and early '80s must have been an exciting time. To Herb Dragert's great satisfaction he had proven his mentor, Tuzo Wilson, right after all. The general public, however, knew almost nothing about it because the latest evidence and the debate it spawned were confined primarily to academic journals, some so specialized that only a handful knew where the cutting edge of this new science could be found.

Even a lot of scientists were unaware of the variety, the volume, and the geographical extent of the data that were piling up. At the annual meeting of the American Geophysical Union in December 1981, more than a few heads turned when John Adams spoke about mountains of the Coast Range tilting to the east. Like Herb Dragert and Jim Savage, Adams was convinced this could only be happening if the two plates were locked together. His presentation at the AGU had an impact on several other scientists.

"I guess something of it caught the eye of the people in San Francisco who were doing the WPPSS project," Adams told me. When he uttered the unfortunate acronym, it sounded like *whoops,* which is the expression almost everyone would come to see as appropriate for the infamous project eventually. The Washington Public Power Supply System was a megaproject involving construction of five nuclear power plants, two of which were to be built in the small town of Satsop, west of Olympia along the mountain highway leading out to Grays Harbor on Washington's west coast.

Even though the nuclear plant at Humboldt Bay in California was already known to be in trouble because of crustal fractures caused by Cascadia's tectonic motion, another pair of reactors were going to be installed pretty much on top of the same subduction zone in Washington State. As luck would have it, several of the geo-engineers from Woodward-Clyde Consultants, the group doing the seismic risk analysis for WPPSS (the same company that had done the assessment of Humboldt Bay), happened to be in the AGU convention hall that December.

As Adams remembered it, "One of them basically said, 'We're interested in what you're doing. Would you like to come down and talk to us?'" The Woodward-Clyde geologist who extended the invitation was David Schwartz. He had met Adams a few years earlier and considered him "one of the more interesting guys in paleoseismology," with a new take on active faults.

The presentation was "for our enlightenment," Schwartz explained. He wanted to stimulate discussion within the consulting team about "what was going on up there" in Washington State in terms of the seismic risk factors that might affect the WPPSS project. Adams arrived at the San Francisco offices of Woodward-Clyde a few weeks later and quickly glanced at a preamble document prepared for the meeting. "I guess you could say it was open-minded," said Adams. "It certainly wasn't coming down very strongly one way or the other." Meaning Schwartz and his colleagues appeared to be scientifically neutral when it came to the question of Cascadia's fault.

The day's agenda included two presentations. Masataka Ando went first and laid out the argument that most geologists at the time believed to be true. As David Schwartz distilled it, "You knew the plate was going down but the question was—*how* was it going down? Was it going down aseismically—in which case you can make the argument you aren't going to have large earthquakes? Or was it locked? That was sort of the crux of the issue." Ando, of course, thought it was aseismic.

John Adams delivered the second presentation, his summary of the Coast Range tilting data. He saved the kicker for the end, wrapping up his talk with an overhead slide of the Griggs and Kulm turbidite landslide data from the Oregon coast. Huge offshore mudslides in deep-sea channels from river systems hundreds of miles apart could have been caused only by very large earthquakes, in his view.

He told the team from Woodward-Clyde, "This is symptomatic of an active subduction zone." He then tried to encourage follow-up research by suggesting, "The earthquakes don't happen more often than every four hundred years or so, but here's the sort of evidence you could use to tie it down." Meaning those mud cores.

David Schwartz felt himself being swayed by the data. "From my perspective even back then, it was hard to get all of the secondary deformation if things were just sliding aseismically," he said. The folding and fracturing of rocks along the shore, the compression in Puget Sound, the tilting of mountains—how could all that deformation happen on the surface if the oceanic plate were sliding smoothly underneath?

Schwartz explained that his company had been hired to perform the FSAR—the Final Safety Analysis Report—on the Satsop power plants, which would soon be presented to the Nuclear Regulatory Commission. Their preliminary draft back in 1974 had been written in a completely different atmosphere, when "the Cascadia Subduction Zone was never a consideration." At that time the official U.S. seismic hazard maps "did not identify it as an earthquake source," explained Schwartz.

The science was changing quickly, however, and so was the political environment surrounding nuclear power plants. In the next draft of the FSAR, circulated internally to the consulting board at WPPSS, a new, more cautionary tone had replaced the earlier optimism. "In that document we were definitely opening the possibility that the subduction zone should be considered as a source of strong ground motion," he told me. This was a very different view of the situation and, not surprisingly, it didn't go over well with the WPPSS officials. They were "not delighted by this turn of events," Schwartz remarked dryly, and things slid downhill from there.

A different firm of consulting engineers had been hired to design the reactor, and they clearly disagreed with this alarmist talk from Woodward-Clyde of potential seismic shocks—Alaska-size jumbo quakes, backed up by a handful of mostly theoretical papers with equivocal conclusions and very few hard facts—which threatened to wreck their chances of building a pair of billion-dollar atomic power stations. They insisted on a face-to-face meeting to put things back into some kind of perspective.

Woodward-Clyde, meantime, had gone ahead, drawn its own conclusions and submitted its final draft of the safety analysis to the Nuclear Regulatory Commission, suggesting that Cascadia might indeed be a problem. Thus the tone of that next meeting was fairly poisonous, according to Schwartz's memory. Two rival groups of engineers faced off against each other with their client—representatives from the Washington Public Power Supply System—clinging to a ringside seat.

"Looking back at this, I have to appreciate the chutzpah they had in coming to our tank and trashing us in front of the client," said Schwartz. "They simply said we were wrong in every assumption we made that the subduction zone could produce large earthquakes, that they could demonstrate that this wasn't the case, that mention of even the possibility of seismic subduction would kill the project with the NRC."

The implication seemed to be that some other company should take over the entire project (removing Woodward-Clyde from the equation) and "obviously this didn't go down well with us," Schwartz continued. "It turned into a year of hell working with those guys!" In 1982 the NRC decided to call in outside consultants to help resolve the issue. They contacted a young geophysicist named Tom Heaton at Caltech and asked him to review the earlier preliminary safety report, the one concluding that Cascadia was *not* likely to generate large quakes.

Heaton told me that after reading the first, more optimistic WPPSS document, he concluded that "the reasoning was weak" and decided to consult a more senior Caltech professor to help him write a response. Dr. Hiroo Kanamori had already done extensive studies of other subduction zones around the world. Kanamori was, in effect, Heaton's mentor and together they had just completed a draft of the paper that would eventually be credited with turning the tide against the aseismic subduction hypothesis.

In essence Heaton and Kanamori compared Cascadia (back then it was still being referred to as the Juan de Fuca Subduction Zone) to all the other active quake-prone subduction zones along the coasts of Chile and Alaska and to the Nankai Trough off the coast of Japan. They found more similarities than differences. Bottom line: if giant ruptures could happen there—in Chile, Alaska, or Japan—the same would probably happen here, in the Pacific Northwest. Their paper, published in the *Bulletin of the Seismological Society of America* in June 1984, argued that the absence of quakes in recent history didn't mean they wouldn't happen in the future.

Like the other seismic danger zones around the Pacific Rim, Juan de Fuca's spreading ridge was so close to the edge of the continent that the new slab of ocean floor pumped out from the bowels of the earth was still relatively warm, thin, and buoyant. Put another way, the seafloor plate had not had much time to cool before it got jammed underneath the landmass of North America. Because it was warm and buoyant, the plate was in all likelihood scraping under the continent at a very shallow, almost horizontal angle. And because it was relatively smooth it was probably sticking to the upper plate the same way wide, flat-surfaced tires known as racing slicks build friction and stick to pavement.

Heaton and Kanamori found that the biggest megathrust events were directly related to young, buoyant plates being strongly coupled to the overlying landmass at shallow angles—which fit the description of the Juan de Fuca Subduction Zone perfectly. All the other extreme danger zones had shallow trenches full of thick sediments that had been piled up against the outer coast and compressed, causing folds, faults, and uplift of the overriding plate. Same with Juan de Fuca.

They found that weakly coupled subduction zones like the Marianas Trench had much steeper angles of dip and much deeper trenches, which caused the down-going oceanic plate to melt sooner and therefore not get stuck to the upper plate. Without being "strongly coupled" to the upper plate, the subduction process does occur smoothly, creeping along aseismically, with no big jolts. But the Juan de Fuca Subduction Zone was nothing like the Marianas Trench. Therefore, Juan de Fuca was probably *not* aseismic.

At least that's how Heaton and Kanamori saw it. Compared to all the other places around the Ring of Fire, the Juan de Fuca zone looked just as capable of deadly temblors as the worst of the lot. "This 500-kilometre gap in seismic activity is one of the most remarkable to be found anywhere in the circum-Pacific seismic belt," they wrote. Their concluding paragraph must have sounded like a call to arms for other geologists: "The Juan de Fuca and North American plates appear to be converging at a rate of between 3 and 4 cm/yr. The Juan de Fuca subduction zone shares many features with other subduction zones that are strongly coupled and capable of producing very large earthquakes. Although the shallow part of this subduction zone shows little present-day seismicity and no significant historical activity, we feel that there is sufficient evidence to warrant further study of the possibility of a great subduction zone earthquake in the Pacific Northwest."

The problem, as Tom Heaton explained it to me, was that he did not have direct physical evidence of earthquakes. All the comparison studies in the world could not prove unequivocally that Cascadia's fault had ruptured in the past. The compression evidence from Jim Savage's survey markers near Seattle, like those squeezed-together mountain-tops Herb Dragert was studying on Vancouver Island, were *symptoms* of stress, not proof that big quakes had already happened here. And that's what Heaton was urging others to find—the proof.

CHAPTER 11

Quake Hunters: Finding Cascadia's Ghost Forest

Fog drifted over the hissing surf and followed us as we paddled in tandem, two canoes across a swollen winter tide up the Copalis River toward a grove of trees that had no business being where they were. In the distance, beyond the glassy green water, stood a thick fringe of marsh grass the color of wheat. Behind that, on higher ground, the dark outlines of heavily timbered hills.

As we got closer I guessed the marsh grass might be waist deep if the ground were solid enough to hold a human's weight. The tan thatch might have been the river's opposite bank in some previous lifetime. Now it was a saltwater maze with a Medusa-like braid of channels meandering through and around it half a dozen different ways to the sea.

The tide marsh looked like a big unmown pasture with these hulking dead cedars scattered across the flat ground, standing in defiance of time, weather, and gravity. How could fully grown cedars several centuries old be standing in knee-deep salt water, their storm-battered trunks naked of bark, bleached gray by the sun and draped in moss and lichens, in the middle of a tidal marsh on Washington's west coast? Western red cedars don't grow in salt water.

One of two things could explain the wrongness of this picture. Either the level of the sea had risen far enough to drown the trees or the land had dropped, slumping far enough below mean high tide to turn a forest meadow into a salt marsh. The two men in the canoe up ahead were about to show us how they had bored and scraped and dug the answer from the damp coastal muck.

Decked out in gumboots, faded orange rain gear, and a green life jacket that was never zipped shut and wearing a bright red tuque against the cool ocean mist, paleogeologist Brian Atwater had packed a folding army shovel and his trusty old Grumman canoe to give us a first-hand look at the Washington coast evidence of monster quakes. The dead cedars had become a vital clue in the ongoing mystery of Cascadia's fault. Atwater's colleague David Yamaguchi, from the University of Washington in Seattle, led the investigation that established the time of death. Together they were a forensic team worthy of their own CSI spinoff.

Atwater's discovery of this ghost forest on the Copalis River in March 1986 did not come about by accident. He'd been driving to the coast at every opportunity for months, specifically in search of proof that subduction earthquakes had left their marks on the Washington shoreline. His journey through the tentacles of saltchuck, sand, and river mud had begun in Seattle in October 1985 when his employer, the U.S. Geological Survey, organized a seminar on seismic hazards in the Pacific Northwest and invited all those doing active research in the region to attend.

If the two plates were sliding past each other smoothly, at a constant rate, and without getting stuck together, then according to Ando and Balazs there should be a slow, continuous, and irreversible rise in land levels along the outer coast. And that was something Atwater figured he could probably measure and verify—or disprove. It sounded like an interesting research project.

On the other hand, if the two plates were *stuck together* by friction, strain would build up in the rocks and the upper plate would bend down along the outer edge and thicken inland, humping upward until the rocks along the fault failed. In the violent, shuddering release of strain during an earthquake, the upper plate would snap back to the west, toward its original shape.

But the clear signal—the geodetic fingerprint—of each individual earthquake would be the abrupt *lowering* of land *behind* the beaches when the upper plate got stretched like taffy and then sank below the tide line as the upper plate snapped back to the west. That's precisely what George Plafker had found in Alaska and Chile. It's also what John Adams thought the mountain-tilting data predicted for Cascadia.

Although journalists and members of the public who attended the Seattle meeting saw little of the ferment and discord behind the scenes, significant doubt and strong disagreement had separated the scientists into opposing camps. "There was plenty of skepticism out there among geophysicists that the zone really was capable of doing this stuff," Atwater confirmed. "There were people saying, 'Oh, there's too much sediment. And there's too much water. The zone has all this high-pressure water that's keeping the fault from sticking tightly.' Or people said, 'Oh, it's too warm to give you big earthquakes.' And then there was quite a bit of discussion about the geodetic evidence of the time—that mountaintops were moving closer together in the Seattle area."

The key study by Jim Savage and his colleagues was hotly debated. Did the data really prove that mountain peaks on opposite sides of Puget Sound were being squeezed closer together? The increments of movement measured were quite small and the Geodolite—the best available technology at the time—was being pushed to the limits of its accuracy. Was the signal Savage detected "robust," or was it just a few spikes in the white noise?

"Savage concluded that the mountaintops were probably moving closer together in the direction of plate convergence," Atwater said, "and he took that as evidence that the subduction zone really is locked.

Some people dismissed it. And others said, 'No, it's a real signal.'"

Atwater told me he decided to focus personally on the vertical motion predicted by Ando. "When they said the Pacific Coast was rising three millimeters a year relative to Puget Sound, I said, 'Ah ha! Three meters per thousand!' I think: You know, those are large amounts—a thirty-foot difference in the level of the shoreline." That was the kind of motion he thought he'd be able to see in the geology. He would go out to the coast and find out whether a three-thousand-year-old shoreline was now thirty feet (9 m) above sea level, simple as that.

Making the debate more interesting, however, were the seismic data that the University of Washington's own Robert Crosson had published more than a decade earlier, in 1972. His fault plane solutions for small to moderate temblors occurring in the upper plate near Seattle suggested a *north–south* compression—directly contradicting Savage's east–west motion. Crosson had concluded that the eastward subduction of the Juan de Fuca plate had stopped.

Although not as widely circulated, there was also the January 1983 PhD thesis of Garry Rogers, who would go on to become a leading seismologist at the Pacific Geoscience Centre in Sidney, British Columbia. Rogers had constructed a theoretical model to explain how there could be north–south ruptures (roughly parallel to the Washington coastline) *and* east–west compression at the same time. He noted that the underlying oceanic plate seemed to be deforming as it *turned a corner* beneath Puget Sound. Instead of continuing toward the east as it had for millions of years, the Juan de Fuca plate had apparently started rotating at an oblique angle to the coast, shifting to a northeasterly movement. As it turned, it created a bulge in the overlying plate that became the Olympic Mountains in Washington State. Stress in the lower plate increased until the rock began to collapse in on itself. This partial collapse of the oceanic slab did not, however, relieve all of the stress caused by making the turn. Resistance to movement along the fault plane caused rocks to shear in some places. Imagine a deck of cards spread horizontally by the force of your hands. Once the stress exceeded the strength of the rock, it fractured in cracks that ran parallel to the coast: the north–south fault planes that Crosson had detected.

Meantime, the continued movement of the lower plate perpendicular to the coast would still cause northeasterly compression, and either the strain would be stored in the rocks until they failed in very large earthquakes or the plates would creep past each other aseismically. The trouble for Rogers and the others who wanted to see this question answered once and for all was that the *seismic* data seemed to suggest the aseismic option. No earthquakes in recorded history.

The only evidence of horizontal strain building up had come from Jim Savage and company. Until further geodetic measurements were completed—a parallel study was already underway by Herb Dragert and his team on Vancouver Island—there would be no resolution. "In the meantime," wrote Rogers in the final line of his cautious conclusion, "large subduction zone earthquakes must be considered a possibility."

Then, to complicate the contradictions, new data were released later in 1983 from Craig Weaver and Stewart Smith's new seismic array, installed after the Mount St. Helens eruption. It revealed a fracture zone in which "maximum compression is northeast, approximately parallel with the direction of plate convergence." Like the Savage data, this was interpreted as "evidence for locked subduction." It seemed a tipping point might soon be reached.

But it was Tom Heaton's talk at the USGS conference in Seattle about the bigger picture—Cascadia's similarities with other dangerous subduction zones—that drew sparks and generated most of the attention. "He was out there voicing some views that were somewhat unpopular," said Atwater. "It did serve as something of a lightning rod." Thus the quest for proof, for convincing evidence, for resolution, became all the more enticing.

The conference made news all over the Pacific Northwest. At the Pacific Geoscience Centre about a week later, when the CBC camera crew and I showed up to shoot our very first Cascadia documentary, Dieter Weichert, who was running the seismology lab at the time, took the opportunity to go public with his own bold statement. He stuck his neck out and told us on camera that Canada's team of experts, just returned from Mexico City, had decided the possibility of a great subduction earthquake was real.

The Geological Survey of Canada was revising the risk assessment for Vancouver and Victoria and the west side of Vancouver Island from fifty–fifty odds of a magnitude 9 disaster, to seventy–thirty *in favor* of a Mexico City- or Alaska- or Chile-type quake.

Brian Atwater drove west from Seattle in March 1986 toward Neah Bay and Cape Flattery, on the northwestern tip of Washington State, and started searching the beaches, tide marshes, and river estuaries for clues about whether the outer coast had risen or dropped. "I went to Neah Bay with Ando and Balazs firmly in mind," he admitted. "I really went out there looking for elevated shorelines." That's not at all what he found.

Neah Bay was as good a place as any to start because the land all around it is so close to sea level it was highly likely he would be able to spot even slight changes in shoreline elevation, no matter which way they went—up or down. To the north, the bay opens onto the Strait of Juan de Fuca. Across the Strait is Canada's Vancouver Island. To the south a narrow green valley runs from the back of the bay all the way down to the Pacific Ocean beaches. Standing there in the cold mist and rain it was easy to imagine how seawater might have filled the entire valley at some point in the past, connecting the Pacific with the Strait, cutting off Cape Flattery from the rest of the Olympic Peninsula and making it an island.

Atwater spent a few rainy days on the marshy floor of this valley. At first he poked holes with a core barrel and came up with nothing unusual, just signs that sand and silt had built the marsh by filling a former bay. No need for an earthquake or even for the chronic uplift of Ando and Balazs to explain this stuff. But late one afternoon, with the tide down, he tried his luck digging into the muddy bank of a stream that emptied into the marsh. Several swipes of his army shovel exposed something odd a few feet below the top of the bank, beneath a layer of sand from the bay.

It was a marsh soil, marked by the remains of a plant he had studied in San Francisco Bay: seaside arrowgrass. Pretty quickly he recognized what he was looking at—evidence that land formerly high enough above the highest tides for plants to be living on it had suddenly dropped down far enough for the plants to be killed by salt water. This subsidence of the landscape had apparently happened very quickly because that uppermost layer of sand (above the peaty soil) had been dumped on top quickly enough to seal off the arrowgrass from the air and keep it from rotting. Which is why the plants were still recognizable hundreds of years later.

This was no gentle or gradual transition zone from one geologic era to another. The peat had a sharp upper boundary caused by an almost instantaneous and probably cataclysmic change in the level of the land and sea. Was it physical proof that the ground had slumped during an earthquake, that the plants of a marsh or forest meadow had been drowned quite suddenly by the incoming tides and possibly buried under the sands of a huge tsunami? Could this finally be Cascadia's real smoking gun?

Geologists would need to ask this question at many more places than Neah Bay to know for sure. A few samples at a single bay would not be enough to prove the case. But Atwater remained there long enough to jot a note.

"I went over to the post office and sent Tom Heaton a postcard," he laughed. "I thought, ah—he's the one person in the universe who'd be especially interested in this result." Instead of the uplifted beach terraces Atwater had expected to find, instead of confirming the Ando and Balazs hypothesis, here he was, mud spattered and dripping, with evidence that Heaton was right.

In April and May 1986, Atwater took day trips from Seattle to other spots on Washington's outer coast. "I visited each of the four big estuaries in southern Washington," he explained. "Copalis, the next one to the south, Grays Harbor, Willapa Bay, and finally the Columbia River down at the Oregon border. And each of these streams had the same signature of abrupt lowering of land and marshes and forests that had been at or above high-tide level, then got abruptly dropped down."

During the summer of 1986 Atwater and two co-workers uncovered evidence of at least six different events—presumably six different earthquakes—that had each caused about three feet or so of down-drop. The distant geographic spacing along the Washington shore could be evidence that the quakes were big. If the coastline had slumped in river mouths and bays that were many miles apart, the quakes must have been big. But it would take further digging along the Oregon coast and up the west side of Vancouver Island to say just how big.

Were they magnitude 8s? Or magnitude 9s? It was too early to tell. Judging by what he'd found thus far in the four widely separated river estuaries, Atwater was pretty sure they were bigger than anything recorded in Washington's written history. To be on the safe side, he still exercised the normal scientific caution with careful wording in what would soon be considered a breakthrough research paper, published in *Science* on May 22, 1987.

"Intertidal mud has buried extensive, well-vegetated lowlands in westernmost Washington at least six times in the past 7,000 years," he declared in his opening line. "Anomalous sheets of sand atop at least three of the buried lowlands suggest that tsunamis resulted from the same events that caused the subsidence. These events may have been great earthquakes from the subduction zone between the Juan de Fuca and North America plates."

May have been . . . Until other scientists read the paper, studied the data, and agreed that Atwater's conclusions were valid, these new discoveries were still not established or accepted as facts beyond a reasonable doubt. The paper would nevertheless create quite a stir. With the possible exception of those turbidite samples recovered by Griggs and Kulm from offshore landslides that also *may have been* triggered by temblors (a hypothesis still doubted at the time), Atwater's sunken peat layers were considered the first direct evidence of large subduction earthquakes in the Pacific Northwest. What happened in Alaska and Chile had happened here too. And would probably happen again.

CHAPTER 12

Cedars, Peat, and Turbidites: A Tipping Point at Monmouth

After more than ten years of polite bickering, Bob Yeats brought them all together—the convinced, the doubters, and the fence sitters—for a brainstorming session to consider whether or not there is a major earthquake hazard in the Pacific Northwest. Yeats, who today is an emeritus professor at Oregon State University, had been lured away from the lucrative trenches of economic geology—where he had worked in the 1950s as an "exploitation engineer" and senior staff geologist for the Shell Oil company in Los Angeles—to the dark mysteries of subduction zones and seismic hazard analysis.

Newly arrived in Oregon, a state with remarkably little in the way of major earthquake history, he counted himself among the fence sitters when the debate about Cascadia's fault began to heat up in the late 1970s. He even joked about it in a book he wrote called *Living with Earthquakes in the Pacific Northwest.* In the introduction he shared his skepticism with others who figured giant temblors were a California affliction. "That was certainly my own view in 1977, when I moved to Corvallis, Oregon, even though I had been studying earthquakes for many years—in California, of course. My neighbor said, 'Earthquakes? Bob, you gotta be kidding!'"

Yeats, however, was in the audience at the AGU convention in 1983 when John Adams delivered his paper about the eastward tilting of the Coast Range mountains. He knew that Adams was pretty convinced the landscape was being deformed by active subduction of the Juan de Fuca plate. The releveling of the highway survey markers proved that the entire block of coastal mountains was tilting and Adams took this to mean that strain was building up.

Adams had also mentioned to anyone who cared to listen his fascination with the turbidite cores found off the Oregon coast in 1971 by Gary Griggs and Hans Nelson, two graduate students working under the direction of OSU professor LaVerne Kulm. Yeats knew Kulm and his students personally, of course, and was well aware of their discovery. Now, here was this young Adams fellow from Cornell, recently transplanted to the Geological Survey of Canada, suggesting the cores might be evidence of very large prehistoric quakes.

Yeats' reactions were doubt and caution. "As a student of natural disasters, I worry about needlessly alarming the public," he wrote in a book aimed primarily at the general reader. "What would be the reaction of people in major cities like Seattle, Tacoma, and Portland to such bad news? 'Cool it, John,' I said. Good scientist that he is, John Adams ignored my advice and published his results anyway."

Not surprisingly Adams' new information made more or less zero impact on the public at large because it was read mostly by other scientists. The media still had not picked up on the story. That was about to change.

Yeats was concerned enough about the implications of a growing body of evidence that he got together with Oregon state geologist Don Hull to organize a special seminar to be held in Monmouth in February 1987, just ahead of the regular meeting of the Oregon Academy of Sciences. The agenda featured John Adams, Tom Heaton, and Brian Atwater, as well as "skeptics who had previously advocated the idea that no earthquake hazard exists on the Cascadia Subduction Zone."

Somehow *The Oregonian* in Portland got wind of the meeting and wanted to send science writer Linda Monroe to cover the story. Yeats was reluctant because press coverage might cause the lead investigators to pull their punches. "I wanted the scientists to be completely candid, not worrying about a front-page doomsday quote in a major newspaper," he explained. Monroe asked him to trust her and he did, so the conference went ahead as planned.

Yeats described the atmosphere as electric going in to the meeting, yet in the end ironically, and perhaps surprisingly, hardly any sparks flew. The presentations were so solid the doubters were either convinced or decided to keep their thoughts to themselves. "There was no argument," Yeats wrote, "no controversy! Most of the scientists at the meeting were so impressed with the results presented by Adams, Heaton, and Atwater that the no-earthquake opposition retreated to the sidelines. The meeting marked a *paradigm change,* a fundamental change in our thinking about earthquakes in the Northwest."

Linda Monroe came away with a scoop. Readers of *The Oregonian* were informed that, in Yeats' words, "Oregon, as well as the rest of the Pacific Northwest, is indeed Earthquake Country! None of us felt as safe after that day as we thought we had been the day before." Oregon had joined the club with California, Alaska, Chile, and the rest.

Before leaving Monmouth, John Adams asked if he could take a first-hand look at those turbidite samples in the core lab on the Oregon State campus in nearby Corvallis. Up to that point he had only read the published reports. What he found when he studied the actual mud cores and data logs apparently strengthened his conviction that a very large series of seismic shocks was the only logical explanation for so many landslides happening simultaneously so far apart along the continental margin. He returned to Ottawa and went to work on a new paper that would put the turbidites front and center as physical evidence of past ruptures on Cascadia's fault.

Brian Atwater, meantime, received word in March 1987 that the final draft of his buried-peat story had been accepted at *Science*. After his talk at Monmouth, excitement about what he'd found at Neah Bay and points south quickly spread. Gary Carver of Humboldt State, who had come to the conference to report his evidence of active thrust faults along the coast of northern California, was one of those impressed by Atwater's presentation at Monmouth.

"I saw it for the first time and it made sense," Carver told me. "So when I came home from that meeting, I flew into the airport there at Humboldt and instead of home, I drove out to the Mad River slough on the north end of Humboldt Bay and walked. The tide was out and I walked onto the tide flat and leaned over the tide channel and reached down there and scraped the bank with my hand and saw the buried peats that were identical to the ones that Brian had been working on. My first thought was—oh, this is part of the subduction zone! That was an awakening moment." Carver chuckled.

He and colleague Bud Burke quickly found and mapped more of the same: layers of peat buried under gray bay mud, some with layers of tsunami sand and the same kinds of buried stumps and forest debris that Atwater had discovered up north. Seven buried marshes would soon be found at Netarts Bay, along Oregon's north coast, by oceanographer Curt Peterson and geologist Mark Darienzo. As many as eight more would soon be found in and around Coos Bay in southern Oregon by Alan Nelson of the USGS and his colleagues. Up in Canada evidence for ten possible tsunamis would be found at the head of the Alberni Inlet on Vancouver Island by John Clague of Simon Fraser University and Peter Bobrowsky of the British Columbia Geological Survey.

While Atwater's 1987 buried-peat paper was still being readied for publication, Tom Heaton saw a preliminary draft and decided to cite the breakthrough in a review article he and Stephen Hartzell, of the USGS, were writing to underline the similarities between Cascadia and other deadly subduction zones. The work begun with Hiroo Kanamori continued with an update in *Science* that in turn got picked up by Walter Sullivan of the *New York Times*.

Sullivan's distillation of this carefully worded warning from one of America's top research labs produced a story with the power to shock any who paid even scant attention. For probably the first time in a nationwide mass-circulation newspaper, the threat posed by Cascadia's fault was given the same kind of serious and sobering treatment as the San Andreas. The opening line was among the least inflammatory yet nonetheless cautionary proclamations written about seismic mayhem.

"Analysis of the geology along the coasts of Washington and Oregon has raised the possibility of an earthquake there as severe as any recorded elsewhere in this century," Sullivan wrote. The meat of Heaton and Hartzell's work was a detailed and specific comparison between Cascadia and the subduction zones of southern Chile, Colombia, and southwestern Japan, which had "repeatedly experienced severe earthquakes." The effects of ruptures like these on the cities of Portland and Seattle would be "difficult to predict, since no modern city has ever experienced such shaking."

In other words, the high impact and long duration of shaking felt in Chile's 1960 magnitude 9.5 disaster has never happened to a large city with a forest of tall buildings. None of the San Andreas temblors has ever hit magnitude 9; San Francisco had only a few tall buildings in 1906; the high-rise core of downtown Los Angeles has never been shaken by a force as strong as a megathrust subduction quake. Even those devastating shockwaves in Japan occurred before most of the high-rise skyline of Tokyo was built. Now two respected seismologists from Caltech and the USGS were warning that Cascadia's fault might do a kind of urban damage never before seen in the modern world.

Sullivan decided to underscore the point that Cascadia could be every bit the menace of San Andreas and then some: "One cause for concern, the authors wrote, is a tendency of earthquake tremors from a descending plate to be far more damaging, at distances beyond 30 miles [50 km], than those from horizontal slippage such as that along California faults. Furthermore, they say, oscillation rates are of a nature especially damaging to large buildings."

Although Sullivan did not mention Mexico City as proof, the images of collapsed apartment blocks certainly flashed into my mind when I read the story. Mexico City may not have been thought of as particularly modern, but its high-rise towers were certainly the focus of harmonic amplification and extreme damage caused by the long wavelength of shockwaves traveling 190 miles (300 km) inland from the 1985 rupture. What Heaton and Hartzell were telling us was that we should imagine the same kind of amplification happening to younger cities with dense clusters of tall buildings that have never been tested in a subduction event—namely Vancouver, Seattle, and Portland.

When the eastward motion of Cascadia's sea floor is added to the westward movement of the North America plate, the "convergence rate is about 13 feet [4 m] per century," Sullivan continued. "According to Dr. Heaton and Dr. Hartzell, the key question is whether the sea floor has been smoothly slipping under the continent or is 'locked' and accumulating strain. If it has been storing elastic energy for a long time, a sequence of several great earthquakes or a single giant one, comparable to that in Chile, would be necessary to relieve the tension."

Even though Cascadia has not produced a major jolt in the Northwest since it was "permanently settled by Europeans in about 1810," Sullivan reported, "there are indications of periodic seafloor landslides and coastal subsidences that could have been triggered by such events in the more distant past." So finally, there it was in print for a general audience to digest: reference in the *New York Times* to the turbidite landslide cores from Griggs and Kulm via John Adams, and to the sunken coastal meadows that Brian Atwater had found. The so-called smoking gun evidence of Cascadia's violent past was now a matter of very public record.

∞

For Brian Atwater the next step was to ask two questions a lot of people were asking him: how big and how often? To find the answers he packed his kit and returned to the coast in the summer of 1987 to conduct a systematic survey by canoe of those three southern Washington estuaries: Copalis River, Grays Harbor, and Willapa Bay. He paddled miles and miles of shoreline and hiked through marsh, muck, and greasy river mud until persistence, and serendipity, paid off again.

In his initial reconnaissance at the Copalis River, he had walked in, venturing only a short distance from the road. "I missed the ghost forest," he smiled, pointing across the lagoon toward the grove of weathered hulks bathed in sea mist and drifting fog. "It continues on upstream for another mile or two. It's spectacular all the way up."

What made the dead trees important was the possibility that they could help pinpoint the year and season of the earthquake that presumably had killed them. The first scientist to try this tactic was David Yamaguchi, who had earned a PhD in forestry from the University of Washington and was working on a project for the USGS to use tree-ring dating as a way of figuring out when Mount St. Helens had erupted *prior* to 1980. He offered to help Atwater by trying this same technique to date the coastal earthquakes.

In May 1987 they took their first trip together to Willapa Bay. Atwater showed Yamaguchi the stumps of Sitka spruce, the main arboreal victims of great Cascadia ruptures. Yamaguchi chainsawed a few samples, but they didn't look very promising because tree-ring scientists usually prefer to sample from tree trunks—not stumps. Unfortunately, the spruce trunks had all but rotted away.

The great moment of good fortune came a few months later when they worked their way through the mist and saw for the first time the weather-beaten and moss-draped trunks of western red cedar—what would become known as the Ghost Forest of the Copalis River. "When

Dave and I first started working together, we didn't know that big forests of dead *cedar* trees existed," Atwater told me. "Red cedar is more durable. The trunks are still here, standing dead three hundred years after they were killed."

They figured similar trunks could probably be found along other tidal streams as well, and the more evidence, the better. Yamaguchi came up with the clever idea of placing ads in coastal newspapers, asking local residents if they knew about any more of these ancient beauties. And they did. Cards and letters arrived pointing them toward ghost forests near Grays Harbor, Willapa Bay, and along the Columbia River, a stretch of the Washington coast nearly sixty miles (100 km) long.

Did they all die during the same year and season? They should have if that entire segment of the coast had broken all at once in a single earthquake. Or did they instead die in different years at difference places as a result of a series of smaller earthquakes? Timing was everything.

Yamaguchi's first effort to establish a time of death for the spruce stumps had failed because, with all the rot, there were not enough rings left to count. But working with red cedars would be different. Step one of the ring-matching process involved finding a group of same-age trees that were at least as old as the ghost forest—and still alive—to establish a baseline growth pattern up to the current date. Wide rings that grow during good years with plenty of rain, for example, should be found in all the trees in the area. The same with narrow rings that grow in years of drought or fires or other kinds of trauma. The patterns should all match year by year, almost like fingerprints of the local climate.

Once this ring pattern was established, Yamaguchi would be able to work backward from the current year's growth ring and assign specific dates to individual rings in the past to determine in which year the ghost forest cedars died. Later in the summer of 1987 he and Atwater found the live trees they needed for comparison. At the time Weyerhaeuser was harvesting the fringes of a stand of old-growth red cedars that had witnessed—and survived—the great earthquake by inhabiting a hillside above tides, on an island in the middle of Willapa Bay.

"It's a shame that these trees were being cut," Yamaguchi commented. "They're beautiful trees. But we recognized that that was a place where we could gather modern reference samples."

Each day after the loggers went home, Yamaguchi cut radial slices from the stumps they had left behind. Back at the lab, he drew diagrams with paper and pencil to confirm that they all shared a similar "bar code" of wide and narrow rings. Then he looked at the samples from the various ghost forests. "A number of them had more than two hundred rings in a series," said Yamaguchi, "and it seemed like, at least theoretically, that would be enough for me to match against the living trees." With a magnifying glass, a sharp pencil poised over his lab notebook, and an apparently infinite supply of patience, he began counting and cross-matching the rings of dead cedars against the baseline of cedars from the recent clearcut.

"When we started coming up with dates," he recalled, "a few of the trees had rings up until the early 1690s." The most precise date he could be sure of was 1691, meaning the trees had lived at least until then— maybe longer. It was impossible to nail the precise year of death because even the durable old cedars had taken a beating after three centuries of harsh coastal storms, bugs, fire, and rot. The outer bark and final ring of growth were simply not there any more. "The earthquake had happened sometime shortly after 1691," Yamaguchi figured, "but I didn't know how many years afterward."

Atwater and others at the University of Washington managed to narrow the timeline a bit more. He returned to the old quake-killed spruce stumps on the Copalis River and at Willapa Bay, where he chainsawed some samples that—unlike the cedar—still had intact bark. At the university, a team of radiocarbon analysts then used the spruce slices to limit the time of death to some time between 1680 and 1720—with a high degree of accuracy. Not an exact date, but at least it was progress.

In October 1987 Gary Carver, Bud Burke, and several of the graduate students at Humboldt State University were putting the final touches on a paper they intended to release in Phoenix at the annual meeting of the Geological Society of America. Their research on the tectonics of northwest California was finally ready for publication. Nothing splashy was planned, just another incremental step along the road to resolution of the Cascadia Subduction Zone mystery.

Bill Israel, a local journalist in Eureka, however, had been paying close attention to the news of this emerging Humboldt County seismic threat and recognized the implications of what the HSU team had found. He knew about Carver and Burke, he had read Heaton and Hartzell's comparison of Cascadia to Chile and Alaska, and he was aware of Brian Atwater's sunken coastline data. In the weeks leading up to the big convention in Phoenix, Israel had been hanging around the geology department at Humboldt State while the Carver and Burke paper was being polished.

Ever since the Ferndale earthquake in June 1975, when cracks had appeared in the concrete roadway leading up to the nuclear power station, Israel had kept tabs on the seismic risk analysis that PG&E was conducting. He knew about the Little Salmon fault and had learned of Gary Carver's growing list of other active fractures. When Tom Collins of the U.S. Forest Service recognized the distinctive rhombohedral fracture pattern in the sandpit across the road from the reactor and filed a report with the Nuclear Regulatory Commission, Bill Israel knew he was on to an important story.

He was also aware that Carver and Burke could not release their results to the media until the paper had been peer reviewed and published or presented formally at a gathering like the GSA meeting in Phoenix. So Israel collected bits and pieces of information, dug up background material about Heaton and Hartzell, Atwater, and others, then bided his time, tacitly agreeing to embargo his story until the morning of the big convention when it would all become official.

"Somehow he got the idea we were predicting a giant magnitude 9 earthquake for Cascadia," Carver told me. If you read the fine print of the document that was released in Phoenix that morning, neither Carver nor Burke said anything about a monster shockwave. But on Wednesday, October 28, 1987, the *Sacramento Bee* carried Bill Israel's story under a headline that proclaimed, "Giant Northwest quake feared." The subhead made it even more ominous: "Researchers say 9.5 temblor possible."

The opening sentence told a story of paradigm shift, another confirmation of the heretical new science: "New geological findings being released today support a growing body of scientific evidence that Northern California and much of the Pacific Northwest may erupt in a giant earthquake, potentially endangering thousands of lives and hundreds of critical structures." The specifics of what Carver and Burke had found were buried seven paragraphs below and on the next page: "According to Carver and Burke, evidence from the Little Salmon fault suggests that the Cascadia subduction zone comes on land at Cape Mendocino, and that earthquakes of 8.5 or greater have occurred on the fault, perhaps over very large areas."

It didn't take long for the *Bee* story to hit the wire services. Before the morning's coffee was cold, it had become a national news item. When Carver and Burke walked in to the convention hall in Phoenix they didn't see the ton of bricks that was about to land on their heads. "I was met by GSA officials and hustled off for interviews with the national press," Carver recalled. "I had no idea what was happening. Bud and I had said nothing about magnitude 9 earthquakes in our talks or abstracts and knew nothing of Israel's article."

Carver was not amused. "The meeting officials put a beeper on me so they could track me down," he said. Through the blur of the next several hours he did "a bunch of interviews" trying to explain the giant temblor story and trying even harder to dispel "the erroneous prediction" that Bill Israel had reported. "It was a trying day," complained Carver. "When I got up to the podium that afternoon to give my talk, my beeper went off. The lecture room was standing room only and there were cameras and lights everywhere. I ignored the beep and got through the talk okay, but my nerves were shattered."

Was the prediction erroneous? Or simply unattributed? A closer reading of the *Bee* story shows that the magnitude 9 line had come from Heaton and Hartzell's paper published back in April. The headline did not specify exactly who "feared" the giant shock, nor which of several groups of researchers mentioned in the story thought a 9.5 temblor was possible. Headlines seldom do. The new data from Carver and Burke were nowhere near as spectacular or unsettling as Heaton and Hartzell's cautionary tale, so the *Bee*'s headline writer simply pulled zingers from the old story to sell a less dramatic new one.

Israel had been hanging around the geology department at HSU long enough to collect the personal quotes he needed to show where Carver and Burke were coming from. "The potential power represented by the magnitude of such earthquakes, Carver said, is 'awesome, incomprehensible,'" Israel wrote. He quoted Bud Burke as saying, "We're living in a major earthquake zone . . . [The fault] is going to go in the next few generations—and it's going to be big. I don't think there's any question about it."

Israel remembers it well. "I may have gotten them in a little trouble, but I don't think they felt badly about what happened at all. In fact, I think it helped propel them," he said. "Gary had done a lot of work in Alaska himself, after the great Alaska earthquake. It was clear at the time that what he was seeing in Alaska was related to what was going on here. So it was a very good time to be a geologist of the new faith. And these guys were all a part of that wave."

Looking back on it today, even Gary Carver sees some irony in the episode at Phoenix. "Interestingly, years later and based on the work of a number of researchers, evidence for great subduction earthquakes has become widely accepted," he told me in an email. Despite the scary headline, Bill Israel's story turned out to be accurate. Scientific opinion eventually caught up with Carver and Burke—even though they never really needed vindication. They had convinced Bill Israel that Cascadia's fault was capable of monster quakes, and Israel did what he had to do as a journalist to relay that dismal news to the rest of the world.

Once scientists had run through all the alternative explanations, eliminating the wrong answers and convincing themselves the new geology was right, then like it or not the next job had to be getting the word out. After OSU's Bob Yeats himself became "a convert," he groaned about how difficult it was to get his wife, his neighbors, and state legislators in Oregon to take the coming quake seriously.

"At first, it was the excitement of a scientific discovery" that kept Yeats motivated. Unfortunately, telling people about a catastrophic seismic threat was a lot like telling them that a lack of exercise and a bad diet and would make them fat. "The reaction was, 'Yes, I know, but I don't want to think about it, let alone do anything about it,'" Yeats wrote in his survival guide to earthquakes in the Pacific Northwest.

After a while, though, public lethargy became a real drag. "Suddenly, earthquake science stopped being fun, and as a scientist, I began to feel like a watchman on the castle walls warning about barbarians at the gate, begging people to take me seriously," he lamented. Despite recurring images in the news of death and destruction from previous subduction disasters, people seemingly didn't have the will to respond. Perhaps burying one's head in the sand is a type of survival mechanism, a way of coping.

CHAPTER 13

Cascadia's Segmented Past: Apocalypse or Decades of Terror?

For Stephanie Fritts, moving to a small town seemed like a logical way to escape the chaos and frenzy of the modern world. When she arrived in her chosen paradise on the western edge of Washington State, nobody said anything about megathrust earthquakes or tsunamis. Ilwaco was an idyllic resort community with a mild climate, white sandy shores, and tall green forests.

She had lived the jet-set life of a fashion buyer for the May Company department store chain, and in the early days it seemed like a great job. Based in Portland, she spent half her life in New York hotels, missing her husband and children way too much. She began to question her "contribution to society" and decided that "clothing people just wasn't doing it."

Then one day her father told her he could use some help running his department store—an old-fashioned local landmark that carried everything from ladies' wear to oakum—in the fishing town of Ilwaco, just across the Oregon line in the lower left-hand corner of Washington State. Here the turbulent outflow of the Columbia River creates Cape Disappointment and the Long Beach Peninsula, a narrow, sandy spit that forms the outer boundary of Willapa Bay. A necklace of quiet little hideaways like Seaview, Long Beach, and Ocean Park were being transformed into retirement and tourist destinations famous for clam digs, sandcastles, kite festivals, and the scenic splendor of the Pacific.

So Stephanie decided to make a lifestyle change. She went to work in her dad's store and not long after settling in started volunteering as an emergency medical technician for the Pacific County ambulance service. She really liked the feel of public service, of doing something positive for society. She had no idea how much bigger that job would eventually become.

Willapa Bay is the main place where Brian Atwater was quietly digging into tide marshes and stream banks in the spring and summer of 1986, finding evidence of huge prehistoric earthquakes and tsunamis. Few local residents were aware at the time that he was in the neighborhood or what he was up to, and it's probably just as well because the news when it finally came out was most unwelcome. Like ships in the fog, he may even have crossed paths with Stephanie Fritts, unaware they would later join forces on a much-needed public safety campaign.

Stephanie's own introduction to plate tectonics and tsunami waves came on May 7 that year, when a distant rupture triggered a chain of events that would rattle nerves and change the lives of people living in Pacific County. On the Wednesday in question her husband, David, who worked in the lumber industry, had driven to Portland on business when a magnitude 8 quake sent shockwaves through the U.S. Naval Air Station on Adak Island, at the far end of the Aleutian archipelago 1,200 miles (1,900 km) southwest of Anchorage.

It was the largest seismic event in Alaska since the Good Friday disaster of 1964. The ground shook and rolled for almost two minutes. Even though the event would be classified as a "great" earthquake (magnitude 8 or higher), it caused only moderate local damage—cracked masonry and concrete walls, collapsed ceilings and partition walls, spalling on concrete beams and piers—all of which was described by the *Anchorage Daily News* as "one of many temblors that rattle the chain every year." Two things, however, made this one different.

The first was that it had been forecast a year earlier by a team of scientists at the University of Colorado. The researchers were looking for precursors, things that change in the earth just before large main shocks. This being one of the most seismically active regions on the planet and the same big subduction zone that had caused the disaster of 1964, a whole slew of new instruments had been installed by 1974—the Central Aleutians Seismic Network—providing a flow of data with enough details to spot even subtle changes.

One of those changes was a sudden drop-off of seismic activity. When all the normal rumble and grind along a big subduction zone mysteriously goes quiet, watch out—something's bound to happen. Or at least that was the theory at the time. When this quiescence was noticed in a segment of the plate boundary near Adak Island, the researchers in Colorado decided to go out on a limb. They said a major earthquake would occur near Adak before the end of October 1985.

When October came and went and the only big rupture was down in Mexico City, the scientists gracefully admitted their mistake and chalked it up to experience as a "failed prediction." Six months later, on May 7, 1986, a quake roughly a hundred miles (160 km) southeast of Adak Island did occur at precisely the location they had thought it would. By then a new computer system had been installed at the Alaska Tsunami Warning Center in Palmer, just north of Anchorage. It was designed to record and locate the focal point of earthquakes quickly, analyze the data, and predict whether or not a tsunami had been generated at the same time.

This was the second thing that made the May 7 shockwave different. It was the first test of the new software. The alarm system was tripped and a team of geophysicists on duty at Palmer had to decide whether to believe the computer. They knew this segment of the Pacific plate had failed before in a magnitude 8.6 temblor in 1957, generating a tsunami that did extensive damage in Hawaii. They confirmed very quickly that the current rupture was definitely located along the sea floor and was therefore certainly *capable* of generating waves. Whether it actually had or not was another question.

Sometimes a slab of sea floor will move more horizontally than it does vertically, so these kinds of jolts don't always lift a wall of water that becomes a tsunami. In the end it was a judgment call. With the odds apparently in favor, the team at Palmer did the cautious thing and notified emergency officials, who sounded the alarm all around the Pacific Rim that a wave had probably been triggered.

From the moment of rupture until the alarm went out only eight minutes elapsed, less than half the time it used to take when the work was done by hand with calculators, rulers, and maps. The new system developed by Palmer station chief Thomas Sokolowski made this the fastest tsunami warning ever issued.

The trouble was that the computer system was still based primarily on seismic data—instrument readings of the earthquake's ground motion—with no quick way of directly measuring waves in the ocean. This was back in the days before deep-ocean detection buoys that could register a change in sea level and provide accurate data about how big the waves were and what to expect in places like Hawaii, Japan, Vancouver Island, or Willapa Bay—where a tsunami *might* be headed. Without the buoys, the new warning system was still an educated guess based on tide gauges and eyewitness reports from coastal communities when and if a wave made landfall.

School was out and the kids had already made it home when the madness began. Up to her arms in soapy water, Stephanie Fritts stood in her driveway and stared at the spectacle for several minutes, not quite sure what was happening. By late afternoon that Wednesday the sun had come out and warmed the day enough to make washing the car a tolerable task, which is what Stephanie was doing when she noticed traffic—lots of it—heading southbound down the Long Beach Peninsula at high speed.

She immediately switched on the radio and heard a cursory news story about a big wave possibly en route from Alaska and wasn't sure what to do next. She tried phoning her husband in Portland. Again and again and again she tried, with no luck. The lines were jammed. She did eventually get through to her parents across town and told them she was on the way to pick them up. She wrangled four children—ages three to fifteen—into the car and decided not to worry for the moment about David because in Portland he should be far enough inland to be beyond the danger zone.

But David was worried about her and the kids. After hearing the same vaguely ominous, fact-free news item on the car radio, he too made a dash for the nearest phone. He called the home number and called and called, unable to get through. So he decided to make a beeline for the family, driving west and north from Portland toward Astoria, where the big bridge crosses the Columbia just below Ilwaco. These were the days before most people had cell phones, so nobody in or near the danger zone could find out what was really happening. For David Fritts it was a shot in the dark.

As the first wave, a 5.8-foot (1.8 m) swell, came ashore on Adak Island, emergency bulletins echoed down the west coast of North America from Alaska to British Columbia and on to Washington, Oregon, California, and Hawaii. The quake had indeed generated a tsunami. In Kodiak a siren started wailing shortly after 5:00 p.m. Alaska time to warn people that a wave could strike there within the hour. Over in Valdez, it was only the luck of the draw that no ships were docked at the Trans-Alaska Pipeline terminal. The captain of an inbound tanker decided to slow his vessel to twelve knots in order to stay in deep water until the danger passed.

Along the Washington shore the U.S. Coast Guard sent airplanes and helicopters equipped with loudspeakers buzzing down the beaches to warn people to head for higher ground. They radioed fishing boats at sea and urged their skippers to head farther offshore. Even on the protected inside waters of Puget Sound the captains of ferry boats were warned away from their docks as the wave approached. A ship laden with dangerous cargo outbound from Seattle toward the Strait of Juan de Fuca was stopped by the Coast Guard and told to wait.

With four children and her parents in the car Stephanie Fritts was finally ready to make a run for it. She started driving southeast toward the Astoria Bridge. Traffic snarled almost immediately. Just outside of Ilwaco, they hit gridlock and were stuck on a country road only a few feet above sea level—waiting for what might be a killer wave—with nowhere to go. High ground was miles away. For nearly 17,500 people in five counties along the coast, a fine spring afternoon had vanished, replaced by a confusing, gut-clenching race to get away from the water as the sun set and the air turned colder.

Eventually Stephanie did get across the bridge at Astoria and as far inland as Westport, Oregon, where she managed to find both high ground and a motel where they could stay for the night. Then things got really crazy. In the darkening chaos of frantic headlights and confusion, David drove right past their motel in the opposite direction on his way home to find them. At the Astoria Bridge the Oregon State Police were allowing traffic to cross the river southbound to escape the Long Beach Peninsula in Washington, but they had blocked all traffic going north. Nobody was allowed back in to the danger zone. David found himself stuck in the Astoria Bridge line-up with no idea where his family was or what was about to happen.

When the first wave from Adak Island finally reached the northern tip of Vancouver Island at Cape Scott, it was only four feet (1.2 m) above mean high tide. When it got down to Neah Bay in Washington, it was only two feet (60 cm) above normal. At Grays Harbor and Willapa

Bay, it was less than that. The maximum height of the Adak tsunami was a harmless 1.8-foot (54 cm) slosh by the time it hit the beaches in Hilo, Hawaii. In Japan it was a hissing five inches (13 cm) of foam. By 10:00 p.m. Pacific time that night emergency officials began lifting the evacuation orders. All that scrambling and racing around in the dark had been for nothing. Stephanie and David Fritts still didn't find each other until later the next day.

Angry and frustrated citizens in dozens of coastal towns called it a false alarm. For others a potential disaster had degenerated into a poor joke. "We gave this big party and nobody came," quipped Lieutenant Commander Tom Pearson of the Coast Guard in Seattle. A train of five small waves were in fact triggered when the ocean floor heaved upward, so in reality it was not a false alarm. There was simply no way to tell how big the waves would be.

For Stephanie Fritts and others with small children to round up and protect, or for those who got separated from loved ones, family, and friends in the panic of those first few hours, the experience was nothing to laugh about. "This was the first I had ever known that the Pacific Coast could be impacted by a tsunami and I was having visions of the old movie *Krakatoa*. I didn't really understand the whole thing and was confounded for the most part! I didn't realize that tsunamis were real and not an invention of the movies," she said.

The thing that stuck in her mind was the lack of information. There had to be a better way of detecting and measuring what happens at sea when a subduction event tears the ocean floor apart. The more immediate and dire implications of an Alaska-type catastrophe happening very close to home, just a few miles offshore from Long Beach—a monstrous quake and train of waves from Cascadia—was another whole movie that nobody had explained to Stephanie and her neighbors. She was annoyed and motivated enough to do something about it.

∞

After the conference at Monmouth, conversations between Brian Atwater and Gary Carver moved into new territory. While others were slowly coming to accept the notion that Cascadia's fault was an active subduction zone, Atwater and Carver and a handful of others had jumped ahead to the next big question. Assuming they had proven the existence of past quakes, how could they find out whether the entire subduction zone had ripped loose all at once—as opposed to rupturing in a series of smaller segments, releasing only part of the accumulated strain each time?

"There were two competing hypotheses," Carver told me. "We were arguing about what *kind* of earthquake it was. Not *whether* it happened." Looking at how far apart the sunken marshes and forest floors were geographically—all the way from Vancouver Island to Cape Mendocino—and sorting through their calendar of radiocarbon dates, they began to speculate about whether or not Cascadia always ruptured from end to end. Was it a single, Alaska-size event each time, or did some segments of the down-going slab—perhaps the southernmost Gorda plate off Humboldt Bay, for example—rupture separately in smaller events? Meaning magnitude 8 disasters instead of magnitude 9 catastrophes.

"There was the Brian Atwater 'apocalyptic model,' which was a magnitude 9 which broke the whole subduction zone," Carver mocked, "and then I had coined the term 'decades of terror' for a series of earthquakes that occurred very close together in time—so when you radiocarbon dated them, you couldn't tell them apart." When not trading quips with Carver, Atwater insisted he was "agnostic" about magnitude 9 and that—all kidding aside—they needed some way to prove *or disprove* the worst scenario.

The decay of carbon atoms gave them dates that were accurate to within a few decades, but that wasn't good enough to tell whether all the marshes and estuaries had been buried at the same time by the same great earthquake and wave. Tree-ring dating would eventually help narrow the timeline, but now there was another wrinkle. The build-up of stress deformation along the coast did not seem to be occurring at the same rate everywhere.

A newer, more accurate releveling survey had been done along U.S. Highway 101 from Crescent City, California, all the way up the coast to Neah Bay, Washington. When Clifton Mitchell and Ray Weldon of the University of Oregon took a closer look at the numbers, they noticed that over the past fifty years the southern end of Oregon and the northwestern tip of Washington had been rising about an inch or more every decade.

The complication was that the area around Newport, about halfway up the Oregon coast, and around Grays Harbor, halfway up the Washington shore, did not appear to be rising at all. To Mitchell and Weldon this suggested that some points along the subduction zone might have *asperities*—rough spots where the two plates tended to hang up or get temporarily locked together. Some parts of the zone—like Newport and Grays Harbor—could be moving without seismic strain build-up while others were locked and loaded for a big rupture. Which sounded like evidence in favor of Carver's "decades of terror" scenario.

Atwater still wasn't sure. He looked at proof of sunken marshes from Vancouver Island all the way down to southern Oregon and possibly including Carver's own tsunami sands near Humboldt Bay in California and thought the magnitude 9 scenario was still viable. He knew he'd have to find some way with tree rings—or whatever—to narrow down the dates. If all those places sank at the same time, it must have been a magnitude 9 or larger and exact dates would prove it.

Carver, on the other hand, could point to the big Nankai Subduction Zone off Japan, which had broken in two magnitude 8 events (1944 and 1946) during and after World War II. If they had been dated with radiocarbon there would have been no way to tell they were separate quakes. Radiocarbon might get you within a decade, but certainly not within two years. Dating the ghost forest's time of death became all the more important.

Here along the west coast, Carver could imagine a sequence similar to the one in Japan. The Gorda plate could break loose first, then the main Juan de Fuca segment a few years later, with the Explorer and Winona segments up north of Vancouver Island after that. Whole decades of terror instead of one big event.

Garry Rogers at the Pacific Geoscience Centre had already written about the possibility of separate ruptures with differing magnitudes. Roy Hyndman and Kelin Wang, also at PGC, were working on a new paper that used temperature variations in the crust to calculate how much of the subduction zone might be stuck and which parts were sliding smoothly due to higher temperatures (near the melting point of rock) deep underground. Their preliminary conclusion was in favor of the magnitude 9 scenario. But soon a new line of offshore evidence from the team at Oregon State seemed to tilt again in favor of the segmented rupture pattern.

With data and new ideas coming together from several directions at once, it was exactly the kind of movable feast that scientists love to sink their teeth into.

In the summer of 1989 a new batch of data from Canada was released, adding weight to what Jim Savage had said earlier about the big squeeze in Puget Sound. Herb Dragert and Mike Lisowski had finally amassed enough evidence from laser beams bouncing off mirrors on the mountain peaks of Vancouver Island to release the numbers at a meeting of the International Association of Geodesy held in Edinburgh, Scotland, on August 3, 1989.

A dozen years' worth of repeated geodetic surveys, along with long-term tidal monitoring along the island coast, had revealed exactly the same kind of squeezed mountains and built-up stress that Savage had seen in the hills on either side of Seattle. Dragert's data confirmed "significant regional strain rates," including a newly detected ridge of uplifted land from Neah Bay, at the northwest tip of the Olympic

Peninsula in Washington, to the mountains northwest of Campbell River on Vancouver Island. The line ran parallel to the subduction zone and was exactly the kind of humping-up you'd expect to see if the ocean floor were stubbed against the continental plate, a big boot kicking it eastward.

The latest seismic data showed all the normal little tremors in the overlying crust and in the down-going plate, but along the subduction interface itself—that eight-hundred-mile (1,300 km) scraping edge where the two plates actually meet deep underground—there was "a distinct absence" of even low-level seismic activity. The same kind of ominous silence that had been noticed before the quake near Adak Island three years earlier.

Herb Dragert's evidence that the subduction zone was locked and loading the kind of strain energy that could only be released in a big shock—the proof that his mentor, Tuzo Wilson, was right about tectonic plates squeezing Vancouver Island against the mainland—was now a matter of public record. The implication was, he wrote, that "a definite potential exists for a future megathrust earthquake."

He and colleague Garry Rogers had also decided that a version of Cascadia's emerging story needed to reach a wider public, so they wrote a short, provocative article entitled, "Could a Megathrust Earthquake Strike Southwestern British Columbia?" which was published in a quarterly government magazine called *Geos*. Hoping, perhaps, that the headline might capture the attention of editors and journalists in the mainstream media, they borrowed the language of a crime novel, talking of "smoking guns" and "lethal weapons."

Piece by piece they laid out the clues unearthed by scientists on both sides of the border. In answering their own provocative questions, they did their best to remain scrupulously neutral while leading readers to the obvious conclusion: "We see no evidence to preclude the occurrence of a megathrust earthquake ... Geological evidence suggests that large earthquakes may have occurred." On one point, however, they did not equivocate; they stated flatly that "crustal deformation is currently taking place."

As for predicting the next rupture, they wrote, "Unfortunately, our present data are too sparse to provide information on the likely timing of such an earthquake in this region." In the closing paragraphs they got serious about the significance of their mystery tale. The consequences of a great subduction jolt and the tsunami it generated would be devastating because so many communities would be affected at the same time. "If such an earthquake occurred, it would likely be the largest economic and social catastrophe due to nature ever to affect Canada," Dragert and Rogers wrote. They carefully noted that much of the evidence was still circumstantial. "As with the pursuit of the smoking gun in the mystery novel, we still need more evidence to prove that the gun has fired or is likely to fire again. But our suspicions are mounting."

The *Geos* article did have the desired effect. It was picked up and rewritten by several newspapers in Canada, making Dragert and Rogers briefly famous as young, intrepid scientific sleuths. Their moment in the limelight was no doubt less painful than the cluster-attack on Gary Carver by reporters at the convention in Phoenix, though the message was essentially the same.

On October 17, 1989, I was back in Toronto, jetlagged from Europe and completely immersed in interview transcripts, trying to write a script for a documentary on Poland's new Solidarity government. I had a quick dinner and went to bed early, feeling exhausted. I had just drifted off to sleep when the phone beside the bed rang loudly and jolted me into a semi-conscious stupor. It was Sally Reardon, one of the senior producers on the desk at *The Journal*, the CBC program I was working for.

"Major earthquake in San Francisco!" she proclaimed, breathlessly.

At 5:04 p.m. Pacific time, during the warm-up for a baseball game in Candlestick Park, a segment of the San Andreas fault broke near the coast deep under a mountain called Loma Prieta, about nine miles (15 km) northeast of the seaside resort of Santa Cruz. For about fifteen seconds the ground shook violently. The crowd in Candlestick Park knew immediately what it was and started moving toward the exits, their barely suppressed panic captured live on network television.

It was game three of the World Series, featuring, ironically, two local teams, the San Francisco Giants and the Oakland Athletics. The first major California temblor covered live on television became known briefly as the World Series quake; geologists would later officially name it the Loma Prieta earthquake. When contact with San Francisco was reestablished, the pictures were disturbing.

Freeways and bridges had collapsed, killing dozens of people. Because the quake happened at rush hour, a traffic helicopter was already in the air and over the water with its camera rolling when a car drifted helplessly in what looked like slow motion over the edge of a fifty-foot (15 m) section of the double-decker Bay Bridge that had dropped like a trap door onto the span below. Because two local teams were playing in the Series, thousands of people had either gone to the stadium or stayed downtown after work to watch the game in bars with friends. As a result, rush hour traffic was lighter than normal. The death toll could have been much higher.

In Oakland things were worse. A mile-long (1.6 km) section of the Nimitz Freeway—built on former marshland—collapsed, crushing cars on the lower level, instantly killing forty-two people and injuring many more. Some cars on the upper deck were tossed around and flipped; others were left dangling over the side as the freeway bucked and twisted during the temblor. Nearby residents and factory workers rushed onto the bridge with ladders, tools, and forklifts and began digging survivors from the rubble. The volunteer rescuers were at it round the clock for the next four days. It would take eleven years to rebuild the freeway.

High-priced houses and condos built on landfill in San Francisco's upscale Marina District shifted on their liquefied foundations. Seven buildings collapsed outright, another sixty-three were damaged beyond repair. Water and gas mains broke, creating hellacious fires. The Embarcadero Freeway, another two-level roadway, was damaged and would have to be torn down. Golden Gate Bridge survived but would need a $75 million retrofit. A six-block stretch of downtown Santa Cruz's historic business district was reduced to rubble.

At magnitude 7.1 this was the largest and most damaging rupture on the San Andreas since 1906. All told, 63 people died that night, 3,757 were injured, and more than 38,000 needed emergency housing. The damage toll topped $6 billion. Once again the world's attention was focused on California's most famous fault.

Meantime, although hardly anyone paid them heed, a small squad of quake hunters continued digging in the mud and turning over rocks from Cape Mendocino to Vancouver Island. While San Andreas dominated the headlines, Cascadia's smoking gun—final convincing evidence of a much larger quake—would soon be found. Quietly and with little fanfare. In tight little rings of western red cedar, in the voices of Native elders who told of a dark and violent night many generations ago, and in sacks of rice destroyed in a shogun's warehouse on the far side of the Pacific.

CHAPTER 14

Digital Water: Catching Waves in a Computer

On a mild spring morning just before lunch, with plenty of California sunshine and a respectable crowd lining the main drag to watch, several normally docile horses spooked as a parade ended. At first the marshals thought it was just flags snapping in the breeze that set them off, but it was more than that. They reined their mounts. A heartbeat later the humans felt it too.

The ground heaved and rumbled beneath their feet. Storefront windows began to rattle and ripple and shatter. All of northern California and much of southern Oregon shook as the edge of the continent sheared away from a down-going slab of the ocean floor. The hilly farmland and redwood forests of the Lost Coast and Cape Mendocino rumbled as the earth rolled and bucked. Lower Humboldt County was ground zero in an event that would make seismic history. The date was April 25, 1992.

When they realized what was happening, the citizens of Ferndale began to scream and run in all directions. Mothers grabbed their children, pulling them off sidewalks and into the street to avoid jagged plates of tumbling glass. Some held straw cowboy hats and sombreros above their heads for protection. People in the street began dropping to a crouch, either because it was too hard to keep standing or simply because they didn't know what else to do. Some fainted or collapsed from fright. Others, cut by glass or grazed by falling cornices and bricks, lay bleeding on the pavement, surrounded by Good Samaritans trying to help. All of this was captured by a television camera crew who had been covering the parade.

Sixty miles (100 km) north in McKinleyville, Professor Lori Dengler of Humboldt State University was getting her family ready for a day of hiking at Patrick's Point State Park, on the beach a little farther up the coast. "Suddenly the ground started to jiggle a little bit," she said, "and then it started to jiggle a lot more strongly." With her very next breath the instincts of a geologist kicked in. "Fortunately by that time I had actually developed a habit of starting to count the duration of an earthquake. It's a very good habit to get into with—one, two," she recounted the beats. "And by the time we got up to about seventy-five, I knew that my plans for the day were completely shot. There was no way we were going on a picnic."

Back in Ferndale the videotape showed piles of splintered gingerbread trim, cornices, and tons of old brick that had crashed to the ground in heaps of rubble and bunting. Thirty-six homes in this Victorian tourist town were seriously damaged. A dozen others twisted off their foundations and collapsed. Forty businesses in a four-block stretch were damaged, putting 80 percent of the town's economic engine on life support.

Pipes broke, gas mains ruptured, and fires started. In nearby Petrolia, population one hundred, the bay door of the fire hall got jammed during the initial shock and was stuck in the closed position. It took several volunteer firefighters considerable time and effort to pry it open before their pumper could respond to the now out-of-control blazes. The post office, a century-old general store, and a gas station burned to the ground. Landslides and rockfalls blocked roads and railway tracks.

As the main shock died away, Lori Dengler made her way outside to begin the next phase of her research. "I lay down on the driveway so that I could feel all the aftershocks," she confessed, apparently unfazed by what this must have looked like to the neighbors. "Earthquakes are really quite delightful if you are in a completely safe place. And I have a very open driveway with no big trees around, so I just lay there for about ten minutes, sort of feeling the music of the spheres—quite literally." She could tell from the duration that the jolt had been at least magnitude 6, possibly higher, and knew it was time to get to work.

Finding the focal point and calculating the strength of the jolt would take several days but a preliminary investigation showed that a nearly horizontal fault began splitting apart six miles (9.5 km) north of Petrolia and seven miles (11 km) underground in a magnitude 7.1 rupture. Little more than twelve hours later, at forty-two minutes past midnight, another quake, magnitude 6.6, centered fifteen miles (24 km) offshore from Cape Mendocino and thirteen miles (21 km) below the surface struck the same general area, causing additional damage. Less than four hours after the second rip it happened again—another 6.6 deep-sea jolt off the Lost Coast. Three strong quakes within fifteen hours. Ferndale, Petrolia, Honeydew, Scotia, Rio Dell—all the small towns at the southern end of Humboldt County took a beating.

The main shock was felt as far south as San Francisco, as far east as Reno and Carson City, Nevada, and across most of southern Oregon. In Sacramento, 202 miles (325 km) to the southeast, a curious thing happened. Lori Dengler did some checking and found that most people living at ground level felt almost nothing. The farther up they were in apartment blocks and condo towers, however, the stronger the motion tended to be.

"If they were in the sixteenth, seventeenth, eighteenth floor of a high-rise building, not only did they feel it—they felt it so sharply that it drove them to run down the stairs and evacuate the building," she said. "By the time you get above the twentieth floor of some of these buildings, more than half the folks evacuated." The astonishing thing was that no physical damage had been done in Sacramento, yet the kind of ground motion generated by the undersea fault off Cape Mendocino was able to travel a long distance inland and cause certain tall buildings to resonate with the frequency of the shockwaves. Shades of Mexico City.

But it was Dengler's cautionary note that really caught my attention. "This was an earthquake that was a thousand times weaker than what we're talking about in terms of the amount of energy in a Cascadia earthquake," she said. Put another way, whenever Cascadia does finally rip loose with a magnitude 9, the results will be off the scale. That's simply the difference between magnitude 7 and magnitude 9.

Several things did, however, make the Mendocino temblor different and indeed significant. It generated a small, three-foot (1 m) tsunami that hit nearby Crescent City, scene of so much destruction and a dozen deaths in 1964. And it lifted up a fourteen-mile (23 km) section of land along the beach at Cape Mendocino. "It turned out that the North American plate that we're sitting on was shoved up and over the Gorda plate, which is subducting beneath us," Dengler explained. In other words, this earthquake was apparently generated by tectonic movement along Cascadia's fault. It was not just another small slip along one of the vertical cracks near the surface.

With a total of ninety-eight people injured and a damage estimate of only $66 million, it may have looked to the outside world like a relatively insignificant lurch compared to the Loma Prieta event on the San Andreas in 1989. There were no dramatic helicopter shots of cars falling through a trap door on the Bay Bridge, no double-decker freeways crushing dozens of cars. But for those living in Humboldt County and those focused on Cascadia, this was a turning point in terms of putting the aseismic subduction argument to rest.

Gary Carver's memory of that deceptively sunny day was just as vivid. The shockwaves hit while he was driving to his office on the Humboldt

State campus in Arcata. After the ground stopped moving he quickly rang home to make sure everything was okay there, then rounded up a bunch of graduate students and headed straight for Petrolia. By about noon the USGS had told them where to find the epicenter. They began scouring the countryside looking at landslides and other physical damage—without ever finding a surface fault.

Because the ocean was high and rough that day Carver and company failed at first to see the quake's most geologically significant water wreckage. Several days after the jolt, however, the HSU crew stopped for lunch in a Petrolia café, where they overheard local residents in the next booth talking about how much the shoreline had changed. Offshore rocks normally submerged were now high and dry and the place "smelled like fish stew."

So Carver and his colleagues headed back toward the beach and sure enough a big swath of California landscape had been hoisted up sharply into new marine terraces. For the next several days the HSU team hiked the shoreline, documented the vertical displacement, and watched while acres and acres of shellfish clinging to rocks that once lay beneath the sea died and rotted.

The seismic data showed an almost flat focal plane and the nearly horizontal motion of a thrust fault, agreeing with what Carver could plainly see on the beach—the upper plate had popped loose from the subducting ocean floor and ridden up over it. "It produced coseismic uplift," he noted, "just like the '64 one did. Except it was all in minia-ture." In essence the continental plate along California's western shore got massively and permanently deformed during the rupture, the same thing George Plafker had seen in both Alaska and Chile.

"When you analyze exactly where that hypocenter or focus of where the earthquake was, it really coincides very closely with where we think that subduction zone interface is," said Dengler, equivocating only slightly. "I mean there's still some debate amongst the scientists as to whether it was really the main subduction zone or a subsidiary fault.

And there's also some debate as to really where the end of the subduction zone is and how complicated things get down there in that Triple Junction region. But I think we all agree that it was a thrust fault and it was clearly related to the subduction zone. And so it became really the first major earthquake to occur on the subduction zone or a very closely related fault."

"The Petrolia earthquake was a subduction zone earthquake because it broke on the subduction zone," said Carver, equivocating not at all. "The boundary between the Gorda plate and the North America plate is a very low-angle thrust fault. And it slipped and caused this uplift and subsidence of the coast and generated this earthquake," he said. "It was a subduction zone earthquake, as far as I'm concerned."

The point of contention seems to be that the Gorda plate, which has broken off the southern end of the Juan de Fuca plate, seems to move and behave separately from the larger slab of the ocean floor and therefore might not be considered a part of the overall subduction zone. And to some extent Carver agreed with this view because evidence of several events found down in the Eel River valley showed radiocarbon dates for Gorda plate ruptures that were completely separate from the Juan de Fuca quakes.

In other words, temblors like the 1992 Petrolia one seem to have happened several times before, with the Gorda plate breaking loose from the overriding continental plate on its own timetable. "It looks like there are some little end pieces that have a life of their own," joked Carver, which to me sounded like yet more evidence for the decades of terror scenario.

The interesting thing was that in Petrolia in 1992 there were three separate ruptures, and to me it looked as if the subduction zone had started coming apart underneath the oil town, working its way outward beyond Cape Mendocino under the ocean floor and toward the main subduction zone. "Once the fault started to unzip, why did it stop?" I asked. "Why didn't it go all the way to Vancouver Island?"

"Yep," said Carver, "that's a very good question. And I've been puzzling on that since '92."

Lori Dengler agreed. "The 1992 earthquake ruptured the southernmost little corner of Cascadia—maybe fifteen miles [24 km] in length. So why did 1992 stop? It's been a long time since the last event. There's been a lot of strain put into the system, so why didn't we have, on April 25th, 1992, a much larger earthquake? We don't have a simple answer to that. Clearly it stopped because it didn't have enough energy to make it through the bump or the asperity or the sticky spot that would have allowed it to go further. Does it mean that we're closer to a larger rupture? Does it mean that we could have another little piece go? Well obviously we're closer. Every day we get a little bit closer."

Proving that Cascadia's fault had started coming apart was only part of the significance of the triple shock that morning in Petrolia. According to Lori Dengler the new data generated by that event caught the eyes of other scientists as well. "What was really important about that earthquake is that it brought two new communities onto the Cascadia bandwagon," she said. "First it brought NOAA. Prior to 1992, the tsunami community really was not engaged in Cascadia."

The National Oceanic and Atmospheric Administration, or NOAA, got interested in Cascadia partly because of the small wave the Petrolia quake shot into the harbor in nearby Crescent City, California. Even though it didn't cause much damage, it did revive painful memories of what had happened in 1964 and it was evidence that a tsunami could be generated by a tectonic source much, much closer to home. Prior to Petrolia the investigation of tsunami damage from Cascadia's fault had been carried out by geologists and seismologists.

Oceanographers and their math whiz colleagues who were attempting to create numerical models of tsunami wave behavior had been concentrating on waves crossing the Pacific from distant sources like Alaska, Japan, or Chile. Their goal was to predict what a given wave would do and devise a better warning system to alert the West Coast. Suddenly it seemed possible that large and damaging waves could be generated within twenty-five miles (40 km) of the beach all along the coastline of the Pacific Northwest.

Seeing what happened off Cape Mendocino, "NOAA became very excited," Dengler recalled, "particularly Eddie Bernard at the Pacific Marine Environmental Laboratory in Seattle. This was an event that really allowed him and his modelers to get their teeth into Cascadia and to actually model that tsunami." Shortly after Petrolia, Eddie Bernard and his wave research team from PMEL were working with state geologists in California to put together a profile of what a larger Cascadia tsunami would look like. Combining a seismic shock and a killer wave in the same scenario had never been done before.

The Petrolia story rang bells in the state capital as well, according to Dengler. "It brought the emergency management community into the picture. Prior to our event, I would say most emergency managers in the state of California weren't really convinced that Cascadia was a problem. They had a really rapid conversion," she declared. "And so we saw an incredible surge in momentum with NOAA and the emergency management community, which really culminated in the planning scenario for an 8.4 earthquake on the Cascadia Subduction Zone."

Dengler paused, looked up, anticipated my next question, then answered it all in the same breath. "Some people say, 'Well, why an 8.4? Why does it *stop* at the California border?' Well, this was funded by the state of California. And so, it's a great document, but it certainly has its limitations." I took this to mean that the mandate for emergency planning by the governor's Office of Emergency Services ends at the California state line. The larger scenario for a magnitude 9 catastrophe along the entire Cascadia margin would have to wait until other state, federal, and provincial governments were sufficiently motivated to get involved. For these other jurisdictions to the north, apparently the tipping point had not been reached yet. The good news was that awareness of Cascadia, along with a new sense of urgency, had now spilled across the boundaries from geology to the liquid sciences as well.

Before Sumatra, very few people had seen a tsunami in action. Until those chilling home videos from Thailand and other fatally ruined vacation resorts were broadcast round the world, hardly anybody in the general public knew what a tsunami could do. Even the experts, oceanographers like Eddie Bernard at PMEL and top-ranked wave modelers like Vasily Titov, had only a theoretical appreciation of the beast they were dealing with. They understood the hydrodynamics, they could do the math and had seen photographs of damage done by waves in the distant past, but until Sumatra neither had seen the real thing in real time.

Before Sumatra, the most recent and memorable tsunami had been the one triggered by the 1964 Alaska earthquake. "I was born in 1962," Titov volunteered, "so I was two years old. I didn't know anything about the tsunami personally. I knew everything from the [scientific] publications and from my studies." Only a few people in affected communities worldwide had ever seen a killer wave and the tsunami scientist who had experienced one was an even greater rarity. Titov set out to solve this problem by capturing a wave in a computer.

He wanted to master the mathematics—and the art—of digital water. If he could learn enough about fluid dynamics to reproduce the behavior of a wave with a numerical model in a computer, he and his colleagues might be able to improve the world's tsunami warning systems and save lives. He recalls how hard it was before Sumatra to get people interested in or even concerned about this rare monster from the deep.

"There was very little awareness in the larger society about the danger of tsunamis," Titov said. "It was difficult to convey this message to society because the first question people would ask is, 'When was the last big tsunami?' And you say, '1964.' It just doesn't sound that convincing." To many the threat seemed as farfetched as getting hit by an asteroid. The work remained an arcane specialty practiced by an elite group of gifted mathematicians who could have held their conventions in a phone booth.

The study of wave mechanics had begun with work on hurricanes and typhoons about twenty-five years earlier. Hurricane science had decent funding because people saw the destructive power of these storms and their waves several times every year. Most of the world's impression of tsunamis was based on scraps of grainy film footage shot decades ago in Hawaii or Japan or on badly faked waves in B-grade Hollywood disaster flicks.

The émigré math whiz Vasily Titov, however, was destined to change all that and NOAA's Eddie Bernard helped make it happen. Titov was one of the new wave of modelers Bernard assigned to the Cascadia problem not long after the Petrolia earthquake. "His models are—the way they convey so much information in such a short amount of time—can only be called art," enthused Bernard, "because in science that's not easy to do."

They first met in 1989 at an international tsunami conference held at the Novosibirsk Institute of Electrical Engineering in Russia—at the geometric center of Eurasia, the world's largest continent. "I remember the banner," said Bernard, picturing the slogan that adorned the meeting hall: "'We are the furthest from any coast in the world, so this is the safest place from tsunamis in the world.' And I think that's true." He laughed, enjoying the irony.

Titov wanted the chance to work with state-of-the-art equipment to develop software that could anticipate the behavior of big waves. "Realizing how dangerous this phenomenon is, we definitely were working on the science of describing the tsunami with the ultimate goal of actually forecasting it," Titov told me. Folding geology, oceanography, and hydrodynamics together in a package that could mine data from several sources at once and then create animated waves that accurately mimic nature in real time was a tall order on a shoestring budget.

So he eventually moved to Seattle, where he joined Eddie Bernard's research team at NOAA's Pacific Marine Environmental Laboratory. Money and technology aside, the odds of getting something like that to work seemed as steep and improbable as forecasting earthquakes or asteroids. Even the best supercomputers back then were struggling to imitate the flow of water. Adding the complexities of gravity and friction across rough surfaces along the bottom, undersea mountain ranges that could steer a moving wave in a new direction, and the infinitely convoluted bathymetry of every harbor and beach—all of which would affect the movement of a tsunami—was a daunting prospect even for someone who loved math. Titov packed his kit and moved from the safest, most tsunami-free zone in the world to one of the most dangerous.

On July 12, 1993, little more than a year after Cascadia's fault started unzipping in Petrolia, California, another powerful Ring of Fire earthquake tore the ocean floor west of Hokkaido in the Sea of Japan, hoisting a mountain of seawater that quickly broke under the force of gravity into a series of tsunamis. On nearby Okushiri Island seismic damage was only one of several disasters. Toppled fuel tanks and broken gas pipes fed fires that spread rapidly through the rubble. Cape Aonae, a peninsula on the south end of the island, was completely overtopped by thirty-foot (10 m) waves. The highest tsunami to hit Okushiri was almost thirty meters—a wall of water nearly a hundred feet high.

The scariest part was that all of this happened in the middle of the night, so people living there never saw the tsunamis coming—yet they clearly knew what to expect. The Japanese had learned enough from painful experience with earthquakes and tsunamis that most of the island's residents instinctively moved to higher ground as soon as the earth started to rumble. Almost two hundred died and many thousands were injured. Homes, businesses, and the main harbor were badly damaged. The toll would have been far worse if more people had lingered in their wrecked villages only to be drowned by the train of towering waves that hissed and roared from the darkness and slammed ashore a short time after the jolt.

There was little that Vasily Titov could do personally for the people of Okushiri Island. By the time he moved to PMEL in Seattle, however, his tsunami model was advanced enough to be ready for a real-world test that might help others in the future. He and his research partners gathered a wealth of new details from the Japanese about where the water went and how high it reached along the beaches. In the tragedy of Okushiri Island there might be just enough new "data points" to fine-tune his and several other models that were being developed so that lives could be saved the next time.

"We cannot say when the next big earthquake is going to happen," said Titov. "However, from the moment a tsunami is generated, if you know some data about the tsunami, our model can actually tell you pretty well what happens next. How high the tsunami wave is going to be at the coastline, how big the impact is going to be at a particular location. The only thing we have to know for that is the measurement of the wave." Not surprisingly, Japanese scientists had made very precise observations of what happened on Okushiri and along the Hokkaido coast.

One of the many tricks to making a computer simulate a tsunami was learning how to create numerical codes that could reproduce the nonlinear movement of water as a tsunami got bigger and bigger. Before the 1993 wave, Titov and others had created several digital simulations that accurately mimicked the behavior of water in laboratory tests. Titov's software even performed well in terms of predicting the outcome of a real tsunami generated in the Aleutian Islands.

"It was not a forecast in the operational sense of the word," Titov conceded, "but I had all the components in my computer. And when the tsunami came, the comparison was so good," he paused, searching for the words, "I could not believe my eyes. In a nutshell it performed much better than expected." The Aleutian tsunami that served as his earliest test case was another of those relatively small waves that caused little damage. He knew that bigger waves were *not* just more of the same. At a certain point they morphed into something else entirely. Two plus two could add up to five or even ten in the nonlinear world of killer waves.

"Tsunamis are such beasts that they change their attitude, if you will, when they grow bigger," Titov explained. "It's sort of a trivial thing to say, but in terms of a mathematical model, it means that it goes from the linear stage to be a nonlinear phenomenon. And nonlinear is much more difficult to predict, much more difficult to model."

The NOAA team needed data from a larger wave to plug in to the computer if they were to see how well the model mimicked what happened when nature went on a rampage. "What was missing was the big event," said Titov, meaning a wave that could "test the system from the beginning to the end." That event—the wave that became the benchmark for his model—was the one that hit Okushiri Island in July 1993. He and colleague Chris Moore took the camera crew and me to an editing room where they showed us the results on a large, high-definition flat-panel screen.

The images mesmerized everyone in the room. There in full 3D relief stood Okushiri Island as the leading wave approached the beach. Instead of black night we could see it all in perfect daylight detail, a view from space that could zoom right down to sea level and hover at any angle to see what the wave would do from every conceivable perspective. Titov and Moore had taken data points from the Japanese scientists, entered those numbers into the computer and rolled the timeline backward to the beginning.

Knowing how the wave ended—how high it pegged the needles on tide gauges, how far up the various beaches it ran—they reverse-engineered the event all the way back to flat water the moment before the Hokkaido quake and could play it again and again by clicking a mouse. This was more than just high-quality 3D animation—they had converted raw data into computer code to recreate the wave, then converted it again to graphical animation files that let them "fly" through the air above Okushiri and look at every hair on this monster's head.

When they hit playback, they ran the event in slow motion to examine exactly how the waves changed shape, size, and direction as they rolled uphill from deep water, scraping across the rough terrain of the foreshore, the fronts of the waves slowing down because of friction and a heavy load of silt and the trailing edges still moving fast, rising high and crashing hard at last against dry land. It was amazing to see, especially when I reminded myself that this was based on real data from the real wave, not the fantasy of some Hollywood special-effects studio.

"What I like about it," said Chris Moore, his hand on the mouse, "is the aerial photograph pasted over so that you can actually see exactly where the town is situated with respect to the wave." It looked like Google Earth come to life in 3D. "This is an airstrip." Moore nodded at the screen. "Each of these little dots is a rooftop—in reds, whites, and blues. So you can sort of see approximately how large the wave is."

Moore pointed to the small peninsula that was about to be overtopped by the tsunami. In the animation a train of three waves approached the beach. "Here it's shallow," he said, hovering the cursor near the southernmost tip of land. "It's deeper water off of here." He moved the cursor farther off the beach. "And this wave front, as it animates through, tends to bend around the headland because the wave is slower in shallower water and faster in deeper water. So it bends right around there."

The computer made it perfectly obvious why the waves would slow and turn as they did. "And then this group of waves here . . ." Moore zoomed closer to a second point of land, the graphics revealing a steep cliff overlooking a small bay. "It also shows reflection off of that headland." The incoming waves bounced off the wall of rocks and ricocheted back across the bay to hit what had been a sheltered cove on the lee side of the incoming tsunami's path.

Suddenly Titov tapped the space bar to pause the wave. "See this kind of fissure when the wave withdrew from the coast and formed a hydraulic shock?" He pointed to a frozen wall of water standing just beyond a beach that had been completely drained of its surf right down to bare sand. "That's the first time—this animation—is the first time I saw anything like that. And if there was no animation, we probably wouldn't have picked it up."

"Yeah," Moore enthused, "let's just single-step through it and see how it goes." He rewound the wave ever so slightly and played it back frame by frame. "So right about here is where it's forming," he mumbled as the leading wave fell back down the beach, taking all the water with it. "So this water is receding just as the next wave is coming in. It's almost forming a standing wave or hydraulic shock."

Because the Okushiri waves struck in darkness there were few eyewitnesses as the tsunami approached. This animation was apparently the only way to know how the shape of the local sea floor had affected the incoming water. Case histories elsewhere explained and verified what the computer was showing.

"There are eyewitness accounts from the Chile event that sounded really weird," Titov offered. "They were saying the second wave was sitting outside—offshore—waiting and gaining force . . . And that's what it was," he pointed again at the screen. "That was this hydraulic trough—not propagating any more, just sitting there . . . The second wave competing with the retreat from the first wave, creating a standing wave pattern."

The point of showing us the playback of the Okushiri tsunami was to illustrate how far wave modeling had come by 1993. "This was in fact the first wave that we've tested our model against in terms of real event simulations," said Titov. "It was the most studied and the largest event before Sumatra, really." The model was still a prototype, however, and much work remained to be done. New research programs were launched to figure out how to simulate the small-scale details of coastal terrain and more complex problems such as the way waves break, how they transport sediment and debris, and how a wall of moving water interacts with solid objects on land.

The lessons of Okushiri were encouraging for Vasily Titov and his colleagues. "You really wonder what's behind it," he mused, "what kind of wonders mathematics can do. Writing equations, putting it in the computer, plotting it. And then you see the wave evolving just like you saw on TV . . . Things that I saw when I did the animation of the tsunami in Japan—we would never see it just looking at the formulas . . . It's really the power of mathematics working for you."

As the research began to accelerate, so did the ramifications of ignoring or neglecting Cascadia as a major public-safety issue. Within months of the Okushiri disaster in Japan, another scare in the Pacific Northwest and sobering new science from Canada would force politicians to take the initial concrete steps toward a viable coastal warning system. Festering debates and old controversies would be put to rest as the scale and potential of Cascadia's fault became glaringly undeniable.

CHAPTER 15

Defining the Zone: Hot Rocks and High Water

Decades of terror, or the magnitude 9 scenario—which will it be for Cascadia's fault? Apparently Gary Carver and Brian Atwater took enough flak from some of their colleagues for using such terms to describe the possible fate of the Pacific Northwest that they decided to tone down the language in subsequent talks. They started referring somewhat in jest to the biggest, full-margin rupture—the magnitude 9 scenario—as a "dinner sausage" earthquake and the series of slightly smaller magnitude 8s as being "breakfast links." The question of which scenario was more likely to happen remained unanswered and vigorously debated.

Early in the new year of 1994, in the AGU's *Journal of Geophysical Research*, a team of scientists at the Geological Survey of Canada took a stab at defining how much of the subduction zone was locked together and which parts might be moving along smoothly. Presumably, if you knew how much of the zone was locked—if there were some way to measure and define it—you might be able to estimate the size of the rupture that would be generated when the thing finally came unstuck. Drawing a line around the "seismogenic zone" would tell emergency planners how close the quake's epicenter was going to be to major urban areas like Vancouver, Victoria, Seattle, and Portland.

The Canadian team gathered and distilled all the latest reports from both sides of the international border that showed how much the outer coast was being lifted up, dropped down, or squeezed together. Then they plotted the boundaries of each type of data to show exactly where the western edge of the continent was being deformed and in what direction. Herb Dragert's laser and GPS surveys were coming to a sharp new focus. Garry Rogers plotted decades of seismic events that showed where all the coastal earthquakes had been and how deep their epicenters were. These overlapping maps of strain and epicentral data points formed the rough outline of the locked subduction zone.

The first thing that jumped out from this updated, multilayered mass of evidence was the idea that the built-up strain and deformation along the western edge of the continent had to be "transient." The strain was obviously getting released from time to time. The rates at which the coastline was being lifted and the mountains tilted—but especially the speed at which the ground was being squeezed horizontally—were considered "geologically unreasonable." If the hoisting and crunching had continued at this rate with no interruptions for a million years, the mountains along the west coast would be several *miles* higher than they are now and would look much like the Himalayas.

The fact that the peaks of southern Vancouver Island, the Olympic Peninsula in Washington, and the Coast Range in Oregon and California are *not* towering piles like Mount Everest could be taken as one more level of proof that the nonstop pressure of subduction must have been released every few hundred years by very large earthquakes. The point of this latest study was to find the edges of the locked part of the zone and figure out how much real estate was likely to slip sideways when the two plates come unstuck next time.

Roy Hyndman and Kelin Wang, coauthors with Dragert and Rogers, had been working for several years on the idea that temperature could tell the story of where these two plates of rock were bonded together by friction and where they were sliding along smoothly. Wang told me that the dangerous part of a subduction zone—the area that can generate an earthquake—cannot be very far below the surface. "It doesn't go very deep," he said, "because when they go deeper and the temperature is too high, the rock becomes too soft to produce earthquakes."

Rocks go from brittle to ductile as they heat up. The brittle ones will grind against each other and friction will lock them together. Hotter, softer rocks won't stick together. The deeper the slab of ocean floor slides during subduction, the warmer and softer the surfaces become. So at a certain depth and temperature, two tectonic plates theoretically would slide past each other without the risk of rupture. For these reasons Hyndman and Wang thought measuring heat flow could show them where the seismogenic danger zone started and stopped.

First they studied published results of laboratory experiments on rock friction and figured out how warm the ocean floor and continental crust would have to be to stick together and get locked. They knew there was a wild card in the equation—something unusual about the Cascadia Subduction Zone—that would complicate their calculations. The accretionary wedge of sand and clay piled along the edge of the fault was getting dragged down with the slice of ocean floor. It was full of seawater. How did that affect the temperature on the surface of the subduction zone? Kelin Wang had a hunch that this sediment played a major role in the size and severity of subduction zone quakes.

"There's a tendency for the rupture to be very long at these subduction zones," he said. Plates can slip along hundreds of miles of rock surface—from the epicenter, ripping along the "strike" of the fault at more than a mile (2–3 km) per second—all in one earthquake. "I think the amount of sediments cause faults like this to be very smooth," Wang continued. "When the fault is smooth, the rupture has a better chance to propagate for a long way." It's not that the sediments lubricate the movement, he clarified. "It just makes the fault property more uniform."

Wang's idea was that instead of the fault having a million little asperities—rough spots, cracks, and jagged edges—the ocean sediment probably coats the surfaces of the two plates, making the point of contact smoother. That way, when stress finally builds up to the point where the rocks fail, long segments fail together.

Realizing that sediment must play an important role, they had to figure out how it affected the overall temperature of the locked subduction zone. The bottom layer—the subducting slab of ocean floor—started out warm because it was so young, having been created relatively recently (in geologic time) by the hot volcanic furnace of the Juan de Fuca Ridge. Wang and Hyndman figured the layer of sediment probably acted as an insulating blanket, trapping heat in the lower plate.

As the ocean slab moved slightly deeper and the sedimentary wedge dried out, the temperature would rise to 300 degrees Fahrenheit (150°C), which laboratory experiments and research at other subduction zones had suggested was usually the minimum necessary to cause two plates to stick together and generate earthquakes. Farther down, where the temperature rose to 660 degrees Fahrenheit (350°C), the lab studies showed that rocks became too soft to stick together. With this in mind, Wang and Hyndman took measurements of heat flow in the earth, marked the starting point where the thermometer topped 300 Fahrenheit and the stopping point at 660 and drew a new set of wavering gridlines on the map of the West Coast.

They found that the "fully locked zone" capable of generating quakes was about forty miles (65 km) wide, running roughly north–south parallel to the coast in the offshore region beneath the outer continental shelf. The inner edge of the transition zone barely reached the western beaches of Vancouver Island. The locked zone was slightly wider in Oregon and almost twice as wide along Washington's Olympic Peninsula.

The report issued by the Canadian team pointed optimistically to the fact that the energy released in Cascadia's next big event would be restricted to this relatively narrow band of rock beneath the continental shelf west of Vancouver Island. "This reduces the expected ground motion and amplitude at the major coastal cities," they wrote. With the city of Vancouver ninety miles (150 km) east of the locked zone, the maximum magnitude of the coming shockwave might also be somewhat less than some pessimists were expecting. So that was the good news.

The bad news was that the rupture would be shallow and therefore have a greater potential to generate large tsunamis. And the long duration (three minutes or more) of shaking would not be reduced by distance from the zone. The quake would still cause major damage, especially to tall buildings, and "events well over magnitude 8 are still possible," according to the report. A separate paper published a short time later by Garry Rogers also reminded readers that being ninety miles away from the locked zone didn't mean cities like Vancouver would get an exemption from Cascadia's effects. "Anchorage is about the same distance from the down-dip end of the Alaska seismogenic zone as Vancouver is from the Cascadia seismogenic zone," he wrote.

The same kinds of cautionary words would also apply to Victoria, Seattle, Tacoma, Portland, and other cities as far south as Sacramento. Yes, the locked zone is a fair distance from the major urban areas—not directly underneath—but don't forget that Mexico City was roughly 125 miles (200 km) away from the 1985 epicenter and look what happened there. So "the zone" had been defined as a swath about forty miles (65 km) wide, locked and ready to rip.

Years later, in 2009, a study led by Timothy Melbourne of Central Washington University would suggest the locked part of the fault was even closer to the big urban areas—within fifty miles (80 km) of Seattle, Tacoma, and Portland. But the question of whether the entire plate boundary would break all at once from end to end remained unanswered. If Cascadia's fault broke today, would it start as a magnitude 8.8 in northern California and continue northward with several more huge quakes over the next decade? Or would it slip all at once in a magnitude 9.2 mega-disaster? To this day, nobody really knows.

For Stephanie Fritts in Pacific County, Washington, the second alarm came eight years after the first. She was working as a volunteer on the ambulance squad on October 4, 1994, when another subduction earthquake (a magnitude 8.2 in the Kuril Islands north of Japan) ripped the seabed of the North Pacific. The Alaska Tsunami Warning Center in Palmer, Alaska, quickly issued a standby alert to emergency officials in Hawaii and to the entire west coast of North America. When the notification hit the desk of Stephanie's boss, Sheriff Jerry Benning, jaws clenched and sparks began to fly.

Sheriff Benning was notified that *if* a wave was generated, it would take six or seven hours to reach the Washington shore. There were still no deep-ocean warning buoys, so there was still no way of knowing how big the wave might be—or even that a wave had been generated for sure—until it passed some place like Hawaii on its way toward North America. Benning clearly did not like having to second-guess the scientists.

Emergency officials all along the west coast monitored the situation nervously for the first several hours. Fritts recalls hearing Sheriff Benning and the county commissioners make a telephone call to some distant island across the Pacific. They were told there had been "no significant sea level change," so they weighed the options, thought about what had happened last time, and decided not to issue an evacuation order.

When officials at the State Capitol in Olympia were informed of the Pacific County decision, they nearly blew a fuse. Two hours before the tsunami was due to arrive, they contacted county headquarters in South Bend and threatened that if they *didn't* issue an evacuation order immediately, the state would. But by that time it was already too late. The 1986 experience had convinced the sheriff and his team that a full evacuation would take at least four or five hours. Evacuations are dangerous. People get hurt. People could die if there was panic.

The Pacific County Emergency Management Council convened an emergency meeting. They were furious that the state government would make what they considered a rash decision at such a late hour. They rang Olympia and told state officials that according to their reading of state law, the authority to evacuate belongs to *local* government. It was Sheriff Benning's call.

Benning resolutely stood his ground—no evacuation order. Meantime thousands of people had heard about the distant quake and possible tsunami on the radio. Confusion and anxiety spread as the wave got closer and closer. Finally, when the tsunami did arrive, it turned out to be only five and a half inches (14 cm) higher than the normal tide.

In early 1996, a year following the second tsunami fiasco, Stephanie Fritts got hired full time to run the Pacific County Emergency Management Agency, a job she still holds today. Her primary day-to-day responsibility is the 911 call dispatch center, but she's also in charge of emergency planning. On day one at work, the Emergency Management Council informed her, "Your number one task is to fix this!"—meaning the tsunami problem. They wanted her to come up with a strategy for dealing with potential killer waves, a rational evacuation plan. At the same time they wanted her to do some digging and find a way to solve the false alarm problem.

Fritts knew next to nothing about tsunamis, so she pulled out a phone book and started contacting people who were already working on the issue. She helped to organize a series of public meetings and invited earthquake program manager Chris Jonientz-Trisler from the Federal Emergency Management Agency to drive out to the coast from her office in Seattle and share what she knew about tsunamis. Complaints about the false or overstated warnings came up at every meeting.

Scores of people, including mayors and local officials, stood up in community halls expressing anger and frustration about the warning system. They wanted answers that nobody really had. The technology available at the time was simply incapable of detecting the size of a wave in the immediate aftermath of an offshore rupture. Jonientz-Trisler told them as much as she knew, based on her experience at FEMA, but the bottom line was that until newer, more sophisticated equipment was developed, pretty much the same confusing and scary uncertainty would spread like a virus down the coast every time a big ocean rupture happened. The citizens of Pacific County made it known that the status quo was simply unacceptable.

By the fall of 1994, Eddie Bernard and his team at NOAA's Pacific Marine Environmental Laboratory in Seattle were quite far along in developing better technology that would eventually reduce tsunami false alarms. Work on the warning system had begun in 1946 after a tsunami generated in the Aleutian Islands devastated Hilo, Hawaii. The original Pacific Tsunami Warning Center was set up in Ewa Beach, Hawaii, and was operational by 1949. Unfortunately there was a 75 percent false alarm rate in the early years.

Upgrades were installed after the 1960 Chilean quake killed dozens in Hawaii and as many as two hundred people in Japan. After the 1964 Alaska event, yet another series of improvements was made and a second warning center was established in Palmer to concentrate on Alaska, British Columbia, and the U.S. west coast states.

The physical detection and measurement of waves in the deep ocean remained the primary technological challenge. While NASA satellites vastly improved the ability to study the atmosphere and the ocean's surface from space—which in turn allowed hurricane scientists to look down from above through the eyes of storms and begin forecasting what big waves and seawater surges would do as they approached the coast—there was still no real-time alarm system that could tell the scientists at NOAA that a seafloor rupture had triggered a tsunami.

Up to that time the two tsunami warning centers depended on data from a network of seismographs to tell them exactly where an earthquake had occurred and what the magnitude was. If they confirmed that the epicenter was under the ocean floor—and if the magnitude was greater than 7—it was entirely possible (but not guaranteed) that a tsunami had been generated. Whether or not to issue a warning was still an educated guess because sometimes the sea floor moves horizontally more than it does vertically. And without the vertical upheaval, a mountain of water does not get lifted and no significant wave is created.

When a tsunami reaches the nearest shore, tide gauges register the sudden sea-level change. By the time the reading is taken, though, it's often too late to issue a warning to nearby residents. So those living closest to the zone that triggers the wave are simply out of luck.

There would, of course, still be plenty of time to warn coastal communities on the far side of the ocean. The waves will take hours to cross the sea. But if those tide gauges closest to the zone did not survive the initial impact—if the equipment was destroyed by the force of the incoming wave—there would still be no measurement of how *big* the tsunami was. Decision makers on duty in the warning centers would not be able to tell people living thousands of miles away what to expect when the waves finally reached them. So the decision to issue a warning, or not, was frequently based on incomplete evidence. NOAA had no hard physical data of its own that confirmed the creation, size, and movement of big waves. The system could not avoid false alarms or the tendency to overstate potential threats.

It was a problem that had to be fixed and Stephanie Fritts wasn't the only one trying to deal with the downstream consequences. Emergency planners in British Columbia and all five Pacific states had already been through enough false alarms to know that evacuations were not only dangerous and disruptive—they were also expensive.

In full-scale evacuations, businesses were suspended. Factories had to be shut down. Restarting complex equipment and production lines always took time and money. In Hawaii, state officials estimated that a single false alarm cost the local economy nearly $60 million. Not surprisingly, they made it known to NOAA and other federal agencies that the need to "confirm potentially destructive tsunami impacts and reduce false alarms" was their top priority. Stephanie Fritts and Sheriff Jerry Benning in Pacific County knew exactly how the Hawaiian authorities felt.

What Eddie Bernard and a team of more than twenty-five PMEL engineers, technicians, and scientists, along with eighty-five partner companies and suppliers, came up with was a four-stage warning system they called a "tsunameter," which does for wave detection what seismographs do for earthquake measurement.

They started with a device that records pressure changes at the bottom of the ocean. Waves whipped up by storms or hurricanes affect only the surface layer of the sea. A subduction earthquake lifts the entire water column from bottom to top. When a mound of seawater several miles deep is lifted and breaks into a series of waves that start to roll across the ocean floor, the weight of all that water can be measured as a change in pressure when the wave passes over the bottom pressure recorder (BPR) developed by the team at PMEL. The BPR had to be able to function under almost twenty thousand feet (6,000 m) of water without needing maintenance for at least two years.

The second stage of the system involved an acoustical transmission device that could beam the pressure data up to a buoy tethered by cable at the surface. This turned out to be the greatest engineering challenge of all, although they eventually found a way to do it. A deep-ocean buoy technology had already been developed for NOAA's Tropical Atmosphere and Ocean weather forecasting system, but the gear needed modifications to make sure it could survive the frequent and more severe storms of the North Pacific.

In the third stage, the buoys would relay the pressure data from the

BPR to an orbiting satellite that would beam the signal back to land. In the fourth and final stage, the data would be received and processed at the two Pacific tsunami warning centers.

That was the plan. Making it happen was something else. They had to build and deploy a new generation of buoys that could withstand an entire year on the wild and turbulent surface of the North Pacific. The equipment for each tsunameter—the BPR, acoustical transmitter, buoy, and satellite relay—cost roughly $250,000, plus another $30,000 per year for maintenance. The most expensive part of the process, however, would be delivering and anchoring the buoy systems in the deep ocean, using ships that cost roughly $22,000 per day to operate.

A prototype to be deployed two hundred miles (320 km) off the coast of Oregon was ready to go by September 1997. It quickly delivered an accurate stream of data, so NOAA decided to install two more. It would take eight different ships on eighteen cruises—more than ninety days of sea time—to set up this initial three-station array. The good news was that it worked better than expected. It was transmitting tsunami data with a reliability factor of 97 percent—much higher than the 80 percent success rate they had hoped for from a prototype.

That was just the beginning. Since the Ring of Fire's subduction zones constantly eat slabs of sea floor, there was an enormous amount of real estate to cover: more than 5,600 miles (9,000 km) of plate boundaries and grinding trenches that could create large earthquakes and trigger tsunamis. NOAA figured they would need buoys spaced 125 to 250 miles (200–400 km) apart to "reliably assess the main energy beam" of a tsunami generated by a magnitude 8 event. Full coverage would require deployment of twenty-five to fifty tsunameter stations.

NOAA, the USGS, FEMA, and the five Pacific states that funded the project realized early on that if and when the network were finally built to its full length, it would still be too small. With buoys this far apart, some smaller but nonetheless destructive tsunamis could slip through the gaps undetected. This floating line of defense between the "tsunamigenic zones" and the vulnerable coastal communities of the Pacific would be permeable.

The engineers and scientists at PMEL went to work on a new generation of satellite communications technology that would work in both directions and on demand. If a moderate-size earthquake ruptured somewhere under the sea and the seismometers picked up the signal, the on-duty crew at the warning centers ought to be able to send out a signal to wake up the tsunameter buoys and get a reading instantly on all the waves coming across the system—even those slightly below the 1.2-inch (3 cm) threshold.

This way they could spot smaller tsunamis immediately and issue— or cancel—warnings based on real data rather than "an absence of triggered data." So the PMEL engineering team continued to plug gaps in the existing buoy system. At the same time, Vasily Titov and his colleagues were fine-tuning their tsunami model software so that it could use the incoming data stream to forecast what the waves would do when they finally made landfall.

Eddie Bernard knew that roughly 900,000 people would be at risk from a fifty-foot (15 m) tsunami, which is what the computer models said might happen if Cascadia ripped apart. Inundation maps were being drawn for California, but work on the Oregon and Washington coastlines had barely begun.

The weakness in the system continued to be the strategic location of the detection equipment. Because the deep-ocean buoys were placed far enough out in the Pacific to give western North America and Hawaii plenty of warning for a tsunami from Japan, the sophisticated new technology is too far away to be of use in a *local* rupture of Cascadia's fault. The buoys are anchored out beyond the subduction zone, so Cascadia's waves would hit the beaches of the Pacific Northwest at almost exactly the same moment that the pressure detectors picked up the signal at mid-ocean and sounded the alarm.

Bottom line: if you're on a beach and the ground starts shaking—and especially if that shaking lasts more than one minute—it's probably a subduction earthquake and there probably will be a tsunami. The shaking is all the warning you're going to get. Head for higher ground immediately and don't wait for any official notification.

Defining Cascadia's zone gave scientists a more accurate sense of what they were dealing with. Building the prototypes for a deep-ocean tsunami alarm system in the Pacific gave emergency responders a way to make better decisions about whether to evacuate coastal communities when a distant fault ripped and heaved the ocean floor. But the potential for megathrust quakes closer to home remained a subject of debate, and the implications of giant waves generated not far off the West Coast had still not sunk in. An air of unreality and deniability hung over the whole business. It would stay that way until somebody could pin a specific date and magnitude on Cascadia's last great rupture.

CHAPTER 16

Cracks, Missing Rings, and Native Voices: Closing In on a Killer Quake

Long before Chris Goldfinger sailed the Indian Ocean in search of mud cores from the Sumatra 2004 earthquake, he dropped a fish off the Oregon coast and found Elvis. He and Bruce Applegate, both graduate students at OSU in the late summer of 1989, went to sea in a research ship called *Wecoma* using side-scan sonar to take state-of-the-art pictures of the ocean floor. The ship towed a chirping metal "fish" at the end of a long cable thousands of feet beneath the sea surface, pinging sound waves off the bottom to create a digital map that looked as realistic as aerial photos showing the terrain of the ocean floor in stunning detail.

Gliding across the wide, flat abyssal plain, the fish kept chirping, sound waves echoed back, and a strange new picture emerged from the deep. In *Living with Earthquakes in the Pacific Northwest*, Bob Yeats described Goldfinger's discovery of a fault that had cracked the floor of the sea channel and buckled the sediments into a low hill. The onscreen sonar image looked remarkably like a man with a guitar, so inevitably it became known as Elvis. Later, of course, the fractured sea lump was formally named the Wecoma fault after the university's research ship.

Over the next weeks and months the OSU team discovered nine more strike-slip fractures off the Washington and Oregon coast: cracks that penetrated both the Juan de Fuca plate and the overriding continental plate. The ocean floor, at the point where it dives beneath the continent, was buckled, crushed, and deformed into cracks and folds very much like the mangled terrain Gary Carver and his colleagues had found onshore around Humboldt Bay, just down the coast in California.

This was literally and figuratively the cutting edge—the point of impact between two tectonic plates. For Chris Goldfinger the bottom line in this wealth of data was that the newfound fractures and deformations in the crust might be telling us something about the width of the locked zone and also about the kinds of rough spots—the asperities—that the down-going oceanic plate could get stuck on, preventing the entire subduction zone from rupturing all at once.

Goldfinger and another colleague, Robert McCaffrey, published their findings in *Science* on February 10, 1995, concluding that a series of "smaller" earthquakes—perhaps magnitude 8s along the subduction zone or even magnitude 7s in these newfound cracks in the upper plate—could account for pretty much all the Cascadia geologic evidence to date. Since nobody really knew how big a quake had to be to drown the tide marshes that Brian Atwater had found, since nobody really knew how big a shockwave had to be to trigger the offshore landslides that John Adams had written about, it was entirely possible that smaller ruptures could have done all the damage discovered on this coast.

With a fractured and buckled outer edge, the North America plate might be *incapable* of magnitude 9s simply because it couldn't build up and store enough strain for a long enough period to generate a full-zone rupture. That was the "good news." The decades of terror scenario seemed to Goldfinger and McCaffrey more likely than a magnitude 9 apocalypse. They suggested that the seismic hazard and public safety implications of Cascadia's fault did not look quite as daunting as they had before.

∞

In the fall of 1995, however, an international team of mud, marsh, and sand diggers thought enough evidence had accumulated to suggest quite the opposite—that a magnitude 9, full-length rupture *had* occurred along Cascadia's fault. And they were willing to speculate that it happened roughly three hundred years ago. A dozen scientists from federal, state, provincial, and university research labs on both sides of the Canada–U.S. border got together and jointly published a summary of all their separate bits and pieces of evidence for Cascadia's most recent quake.

From John Clague and Peter Bobrowsky's samples of dead plants from sunken marshes that had been quickly covered by sheets of sand left behind by tsunami waves sweeping across the western beaches of Vancouver Island near Tofino and Ucluelet, to Gary Carver's similar evidence of drowned trees in northern California, the picture looked remarkably consistent all the way down the coast. What Brian Atwater had found in estuaries along the Washington shore, Alan Nelson and his USGS colleagues had found in Oregon. The dozen scientists spent considerable effort—including eighty-five new radiocarbon-dated samples—to obtain the most accurate timeline possible. They found that all the ghost forests and marsh plants had been killed at roughly the same time as the land dropped down and was covered by tsunami sand—roughly three centuries ago.

Given the long distance between Tofino, British Columbia, and Humboldt County, California, the dozen "marsh jerks" (as they jokingly called themselves after the jerky spikes in a sawtooth graph denoting the quake-induced sinking of land) said it was all Cascadia's fault. The plate boundary was the only fault common to all the far-flung sites. They reported their findings in a paper published in *Nature* in November 1995. But because there was still reasonable doubt about the exact dates, they still had to equivocate about whether all these events had occurred at exactly the same moment. The paper concluded that a single magnitude 9 earthquake, "or a series of lesser earthquakes," had ruptured most of the length of the Cascadia Subduction Zone "between the late 1600s and early 1800s, and probably in the early 1700s."

Any one of these studies viewed in isolation might not have been enough to convince the most stubborn skeptics. Taken as a whole, however, and seeing that they all said basically the same thing, this united front of twelve top-level scientists looked like a critical mass. Having to equivocate a bit by including the phrase "or a series of lesser earthquakes" no doubt rankled those who wanted to make the clearest, least ambiguous statement possible. But counting the slow decay of carbon atoms to find out when something happened hundreds of years ago was just too imprecise. And there were many examples elsewhere of several earthquakes occurring in series, several years apart. So this remained an unsolved mystery and a real debate.

What they still needed was a more precise date and some way to show, convincingly, how big that mega-shockwave had been. If they could say that an earthquake happened in a specific year, or better yet on a specific day, the whole thing would become more real, more believable not only to skeptics in the science community but to elected officials, emergency planners, and the people who live within striking distance of Cascadia's fault.

In all the mucking about along the coast, scientists had become increasingly aware that people had in fact been living on the edge of Cascadia's rupture zone when it tore itself apart the last time. Beneath a thick layer of tidal mud in the Nehalem and Salmon River estuaries of northern Oregon, Rick Minor of Heritage Research Associates in Eugene and Wendy Grant of the USGS found fire pits full of charcoal and woven cedar mats. Near Willapa Bay in Washington State, fishing weirs and cobblestones modified by fire were found buried in submerged shorelines. Trying to figure out when the campfires had been doused or when coastal villages and their Aboriginal residents had been shaken from their sleep or perhaps drowned by killer waves turned out to be exceedingly difficult.

Anthropologists had been collecting flood and disaster stories from tribal elders since the 1850s but it was only in the 1980s that they began to compare notes with geologists, who found the oral histories intriguing and frustrating at the same time. Judge James Swan, who had lived among the Makah people of Neah Bay, Washington, for a time in the late 1860s published the family history of Billy Balch, a Makah leader, who told him about a catastrophic flood that had turned Cape Flattery into an island. The Balch story began "a long time ago" although "not at a very remote period," when the water of the Pacific flowed like a swollen river through the swamp and prairie between his village and Neah Bay.

This, of course, was the same tide marsh where Brian Atwater found his first evidence of land subsidence caused by large seismic ruptures. Balch's story does not mention the ground shaking, so perhaps the flooding he talks about was caused by a distant tsunami from across the Pacific. In any case, Balch told Swan that after the initial flooding, the ocean began to recede and left Neah Bay dry for four days. It then rose again "without any swell or waves" and submerged the whole of Cape Flattery.

"As the water rose, those who had canoes put their effects into them and floated off with the current which set strong to the north," Swan paraphrased Balch. "Some drifted one way and some another; and when the waters again resumed their accustomed level, a portion of the tribe found themselves beyond Nootka [on Vancouver Island] where their descendants now reside . . . Many canoes came down in the trees and were destroyed, and numerous lives were lost."

"Could this be an account of a great tsunami?" asked Tom Heaton of the USGS in a paper he wrote with Parke Snavely in October 1985. "Great tsunamis may have periods of tens of minutes to hours [between waves] but 4 days is without precedent. If the event is real, then it is apparent that the effects must have been substantial. However, it seems incredulous that any tsunami could have overtopped the entire Cape Flattery region since the highest elevation of the cape exceeds 400m."

Careful not to dismiss the story as myth, Heaton and Snavely tried to imagine what sort of geological event might explain it. The tsunami from the 1964 Alaska quake, which did so much damage to Port Alberni and Crescent City, delivered to Neah Bay a wave that was only 4.3 feet (1.3 m) above the tide, so a tsunami that could completely submerge the entire cape seemed a remote possibility. Then they wondered—what if the land behind Neah Bay had been hoisted up during a *seismic* event, draining sea water from the bay and leaving the land temporarily dry? Heaton cautiously wrote that "crustal deformation associated with a nearby subduction earthquake could explain the uplift. However, this and any other conjectures about the significance of this legend are purely speculative."

If there was a quake, why didn't Balch mention the shaking when he told the story to Swan? Living as close as they did to the subduction zone, the Makah people must have experienced the worst imaginable ground motion. This doesn't sound like the kind of detail a person could easily forget. Heaton and Snavely found enough inconsistencies to consider that the story might be "entirely fictional." In the end, however, they decided it was "noteworthy that such a report exists for a region where there is growing concern that large subduction earthquakes and subsequent tsunamis may be a real possibility."

Farther down the coast, along the beaches of southern Oregon and northern California, Deborah Carver gathered stories told nearly a hundred years ago by Wiyot, Yurok, Tolowa, and Chetco people. From six different villages along at least two hundred miles (320 km) of coastline came similar cultural memories of a violent rupture that struck at night followed by many aftershocks and tsunami waves that rolled in for hours. Many people were killed as villages were inundated or swept away. By the light of day the survivors found places where the ground had liquefied and slumped. In a Yurok village near what is now Redwood National Park, south of Crescent City, Carver learned that one purpose for the ceremonial "jumping dance" was to repair or relevel the earth after an earthquake.

As geologists continued their investigations, more and more stories like this came to light, including one that I heard in Pachena Bay on the west side of Vancouver Island. Seismologist Garry Rogers of the Pacific Geoscience Centre had told me about a reference he'd found in the provincial archives to an earthquake that rocked villages along the west coast of the island in the middle of a winter's night long ago. It didn't take long to find an elder who knew the details, a descendant of the people who once lived in Pachena Bay. The year was 1998 and I was filming another documentary for the CBC called *Quake Hunters*.

Crossing the rocky spine of the island westbound, I thought about living on the edge of the known world two or three centuries ago when the earth tore itself apart. *Wild, remote,* and *spectacular* were words that came to mind. Not for the faint of heart. Since getting from the sheltered eastern shore at Nanaimo to the wild west coast presents a challenge even today, imagine what it was like back then.

After a fast and curvy two-lane tour along the postcard edge of Cameron Lake and through the ancient cedar tunnel of Cathedral Grove, the slow grind up and down the switchbacks of Mount Arrowsmith and the deeply rutted trail beyond Port Alberni became more work than fun. Potholed and scattered with fallen rock, the unpaved washboard logging road that runs parallel to the fjord out to the fishing village of Bamfield and Pachena Bay was a challenge. But once the camera started rolling, we knew the ride had been worth the effort. We were closing in on a story about Cascadia's last big quake.

Around a crackling fire in a big stone hearth in a house just above the beach sat half a dozen young children, mesmerized by one of their elders. Outside, pounding surf and a fine mist of rain washed against the dark green shore. Sitting on a spattered windowsill, his back to the churning sea, Robert Dennis, chief councilor of the Huu-ay-aht Nation, waited until all the young eyes were trained on his weather-beaten face and then began his story in a low, quiet voice that forced the children to sit forward and listen carefully.

"The story that I'm about to tell happened a long, long time ago and was told to us by Chief Louis Nookmis," said Dennis, grandson of Nookmis, the man who originally told this story to an anthropologist back in 1964. "Some of you are directly related to that man," the chief added, nodding and calling the names of several children at his feet.

Like the ancient cedars of the ghost forest in Willapa Bay, the Huu-ay-aht people felt a violent shudder in the earth one dark and stormy winter night many generations ago. "They had gone to bed," said Chief Dennis, "just like any other normal night. Had gone to bed, gone to sleep. And they were awoken during the night when the earth began to shake." He paused, letting the words sink in. "The earth shook. Startled the people. They woke up." He shrugged very slightly. "Thinking every-thing was over, they just relaxed. A little while later, the water came in real fast. Swept their homes away. Swept everything away." The chil-dren were transfixed. "The water just came too fast. They didn't have time to go to their canoes. So all the people that were living there were drowned. They were all wiped out."

Chief Dennis let the thought sink in and then explained how it was that the story itself survived the death of an entire village. The modern fishing village of Bamfield is next door to the rebuilt Native commu-nity where Chief Dennis and his people now live. On slightly higher ground, however, in the trees behind the beach, was another settlement of people who watched what happened that horrible night long ago and who did live to tell the tale.

"They who lived at 'House-Up-Against-Hill' the wave did not reach because they were on high ground . . . Because of that they came out alive. They did not drift out to sea with the others." This according to the 1964 transcript of the story told by Chief Nookmis. For scientists eager to follow the story back in time, the original translation was a bit garbled, making it impossible to figure out when this disaster might have happened. Years later Robert Dennis would help organize a more thorough translation that narrowed the timeframe somewhat. The newer, more detailed version placed the event some time between 1640 and 1740, the same period in which geologists had pegged the last megathrust earthquake on Cascadia's fault.

Brian Atwater is fond of saying that "the earthquakes had written their own history" by dropping segments of the coastline a meter or two, by forcing sandy seawater across the freshly sunken land, and by causing the edge of the earth to crack. The problem was that by the mid-1990s researchers had reached an impasse in their attempts to define how large the quakes had been. The answer to this question was important because emergency planners and civil engineers needed to know what they were up against when it came to designing new buildings and reinforcing old ones. "How great an earthquake should a school or hospital be designed to withstand? How large a tsunami should govern evacuation plans on the coast?" asked Atwater in a book he cowrote several years later.

Atwater's colleague David Yamaguchi had come closer than anyone else to nailing a more specific date. He went back to the ghost forest at Copalis River several times in search of wood cores that might contain enough growth rings to reveal what year the trees had died. He put together a database, a pattern of rings collected from twenty-three ancient cedars that had been alive at the time of Cascadia's last violent outburst. Like the Native people who lived above Pachena Bay on Vancouver Island, the trees of Washington State stood on slightly higher ground and had witnessed the disaster unfolding below.

When they compared the rings of "witness trees" with rings from the ghost forest, they knew the forest floor had sunk beneath the tides some time in the late seventeenth century. "The closest I could get was 1691 as a limiting date for the earthquake," Yamaguchi told me. "The earthquake had happened some time shortly after then, but I didn't know how many years afterward." Insects, rot, and roughly three hundred North Pacific winters had pitted and peeled away too much of the outer bark of these standing dead cedars. The final few years of growth rings—the crucial last clues—were missing.

A dating technology was being tested at the University of Washington, however, and it might pick up where the tree rings left off. Atwater was keen to try it. "Brian realized that he could use the preliminary 1691 date as a tool to help him do something called high-precision radiocarbon dating," Yamaguchi continued. Wood samples were locked for weeks inside shielded atom counters and the resulting timeline was accurate to "the 95 percent confidence interval."

Like the weather-beaten tree trunks, though, radiocarbon technology fell slightly short of the mark. It moved the goalposts closer together without truly scoring a victory. The window had shrunk a little more, putting Cascadia's most recent event some time between 1690 and 1720, but there was still no certainty about whether a single, giant rupture or a "swift series of merely great" earthquakes had done the job. Nevertheless, when news of this tighter timeline became public, it created a buzz. And a scientist from Japan had one of those eureka moments that researchers everywhere dream of.

CHAPTER 17

The Orphan Tsunami: Final Proof of Cascadia's Last Rupture

At a science conference in Marshall, California, in September 1994, Kenji Satake was having lunch with Alan Nelson of the USGS when Nelson mentioned the frustration he and others were having with radiocarbon dating. Nelson and eleven colleagues were putting the final touches on a paper that was meant to collect and focus all the physical evidence for past quakes on Cascadia's fault in a single volume. Satake's specialty was subduction zones and tsunamis. He had studied the effects of the Okushiri disaster in Japan the previous year and was very interested to hear the latest news about Cascadia. Especially Atwater and Yamaguchi's work with tree rings.

"I did not know much about the calibration technique of radiocarbon dating using tree-ring records," Satake told me. "I met Alan. He explained to me the basics of the technique and I was very much impressed by the small uncertainty—only a few decades." Something about those dates—a time span from 1690 to 1720 for Cascadia's last rupture—caused Satake to pay special attention to this latest study.

He was working at the University of Michigan as an assistant professor. Before that he'd completed a two-year term as postdoctoral

In San Francisco, on Market Street near the Ferry building, the ground beneath the street collapsed. This photo was taken on April 20, 1906, two days after the quake. (*U.S. Geological Survey*)

A scene of near-total destruction in downtown San Francisco on May 7, 1906. The outer framework of the California Hotel (center) survived the quake, while all around it was rubble and ruin. (*U.S. Geological Survey*)

A scene from the Alaska earthquake of March 27, 1964. Fourth Avenue in Anchorage collapsed when the ground subsided eleven feet (3.3 m) vertically and lurched fourteen feet (4.3 m) horizontally at the same time. (*U.S. Geological Survey*)

A man examines a ten-ply tire through which a plank of wood has been driven by a wave—an indication of the violent force of the tsunami surge that struck Whittier, Alaska, on March 27, 1964. (*U.S. Geological Survey*)

This photograph, taken in Port Alberni, British Columbia, on March 28, 1964, shows the aftermath of the Alaska earthquake and tsunami. Two of six waves that travelled more than 1,800 miles (3,000 km) from the Gulf of Alaska roared up the Alberni Inlet, flipping cars, smashing fifty-eight homes, and damaging 375 others. (*Alberni Valley Museum Photograph Collection, PN13805/Charles Tebby*)

The Alaska tsunami of 1964 also hit Crescent City, California, where more than a dozen people died when they ventured back into the danger zone after the first wave, thinking the worst was over. Four of the six huge waves generated in Alaska struck the California coast with deadly effect. (*Del Norte County Historical Society*)

An aerial view of the eruption of Mount St. Helens, Washington, near the Oregon border, on May 18, 1980. This photo shows physical evidence of a tectonic plate being shoved beneath North America along the Cascadia Subduction Zone. (*U.S. Geological Survey*)

Earthquake energy traveled long distance from the West Coast to Mexico City on September 19, 1985, with devastating effects on high-rise buildings, which vibrated in harmonic resonance with the low-frequency shockwaves. (*U.S. Geological Survey*)

Paleoseismologist Brian Atwater discovered this ghost forest on the Copalis River in Washington State in March 1986. The trees and other freshwater plants were killed by salt water when coastal lowlands dropped below high-tide level, probably during Cascadia's last megathrust earthquake on January 26, 1700. (*U.S. Geological Survey/Brian Atwater*)

The Loma Prieta (or World Series) earthquake of October 17, 1989, caused deadly destruction in the San Francisco Bay Area. This photograph shows the support columns that failed, causing the collapse of the Cypress Viaduct (Interstate 880) in Oakland. (*U.S. Geological Survey*)

This photograph from March 2007 shows signs of a new public awareness of tsunami dangers on West Coast beaches. Residents of Seaside, Oregon, have learned the tsunami evacuation route and know they will have as little as fifteen minutes to leave the downtown core once a large quake on the nearby Cascadia Subduction Zone has begun. (*Doug Trent*)

Hauling a turbidite core sample aboard the research ship *Roger Revelle* off the coast of Sumatra in May 2007. Offshore landslides triggered by massive earthquakes have left a series of "tectonic fingerprints" in the deep-sea mud. (*Chris Aikenhead*)

A tsunami simulation at Oregon State University in August 2007. Using a scale model of the resort community of Seaside, the simulation shows how a thirty-foot (10m) tsunami expected from a Cascadia Subduction Zone earthquake would sweep across the community, inundating nearly all the downtown business district and many residential areas. (*Omni Film Productions Ltd.*)

In August 2007 cinematographer Ian Kerr (left) prepares to record a tsunami simulation with a high-speed camera (protected by a plastic bag). The snorkel lens will provide a graphic, street-level view as the wave rips across this scale model of Seaside, Oregon. (*Omni Film Productions Ltd./Scott Spiker*)

A tsunami simulation in Seaside, Oregon, in August 2007. The wooden blocks represent residential dwellings, single-story commercial buildings, and multi-story condominiums or hotels. (*Omni Film Productions Ltd./ Scott Spiker*)

Cascadia's wave makes landfall on the west coast of Vancouver Island in a computer-generated tsunami simulation created in 2008 for the documentary *Shockwave*. The first of several tsunami waves can be expected to hit areas from British Columbia to California fifteen to twenty minutes after the earthquake. (*Omni Film Productions Ltd./Artifex Studios*)

A computer-generated illustration from the documentary *Shockwave* (2008) shows the leading edge of a tsunami generated by a megathrust earthquake along Cascadia's fault. The tsunami arrives in Ucluelet harbor on the west coast of Vancouver Island fifteen to twenty minutes after the shaking stops. (*Omni Film Productions Ltd./Artifex Studios*)

researcher at Caltech, where he had studied Cascadia's quake and tsunami potential with Tom Heaton. Heaton, Hiroo Kanamori, and Stephen Hartzell were then banging the drum about similarities between Cascadia and Chile and other big subduction zones. But in 1994 there was still not enough *unequivocal* evidence of past ruptures in the Pacific Northwest for the case to be ironclad.

Later that same year Satake made a point of meeting Brian Atwater on a field trip to the coast as part of the Geological Society of America's fall meeting. Satake would soon depart the United States for a job with the Geological Survey of Japan and was keen to see the Cascadia evidence first-hand before he left. Curiosity piqued by Atwater, Satake wanted to see more.

"He spent some time with me, down in Humboldt County," confirmed Gary Carver. His recollection of events was that Satake zeroed in right away on the dates. "He latched on to that 1700 plus or minus a few years—ten years was the range—that we thought was statistically important." But what about 1700 stuck out in Satake's mind? He had a hunch and followed it all the way home.

When I met Kenji Satake in a small city on the southeast coast of Japan several years later to interview him, he explained why the date had jumped out at him. "If the earthquake happened in Cascadia, and if it was large enough, it must have generated a tsunami which would propagate across the Pacific. And 1700 is not very long ago for Japan. So if such a tsunami arrived in Japan and caused damage that *must* have been documented in historic documents."

No such written record existed in North America. The only cultural evidence of a huge earthquake around 1700 would be the oral histories of the Yurok or the Makah or the Huu-ay-aht people of Pachena Bay. And their stories did not include exact dates. By then a few European explorers had come and gone but the rest of the world knew little about the northwest coast of America and nothing at all about its earthquakes. Even the geography was a yellowing parchment void on most existing maps. The Spanish had outposts in Chile, Peru, Mexico, and southern California, but Cascadia was still *terra incognita*.

By comparison the Japan of 1700 already had a long history of writing things down. The damage reports of earthquakes, tsunamis, and volcanic eruptions were usually kept at temples or in the ledgers of local merchants and public officials. When Kenji Satake returned in 1995, he started looking for catalogs compiled by various teachers and scholars who had been working on a history of ruptures and tsunamis going back more than a thousand years. He knew that some of the tsunamis recorded would have no earthquake listed alongside them because the plate motion that caused the waves would have happened so far away—in places like Chile or Alaska or Siberia's Kamchatka Peninsula—that local residents would not have felt the shaking. Without "parent" earthquakes, the "orphan" tsunamis should stand out and be relatively easy to spot in the seismologists' catalog.

He approached Kunihiko Shimazaki, Yoshinobu Tsuji, and Kazue Ueda, colleagues who had begun studying Japan's historical earthquakes and tsunamis in the 1970s. Together they searched backward chronologically through the list. "Soon after we started looking," Satake said, "we found that there were some documents describing a tsunami from an unknown origin in the year 1700." A series of waves swept down the Japanese coast after midnight on January 27 and into the early morning hours of January 28, to be exact.

"The people studying the history of Japanese earthquakes—for them it was a strange event. There wasn't an earthquake, so they didn't care," Satake explained. "They didn't care what the origin was because they were interested only in Japanese earthquakes—and this wasn't a Japanese earthquake."

While this may have seemed like a eureka moment, plenty of work remained before Satake could say for sure this wave of unknown origin had come from Cascadia. Perhaps it was caused by a typhoon or some other big storm. Maybe the orphan wave had come from a rupture in

Alaska or Chile. So Satake and his colleagues quickly moved to the next stage of their investigation.

Fairly easily they ruled out the storm surge idea. "We knew that the tsunami waves were documented in many places along the Japanese coast," Satake continued, pointing to a map that showed nearly 560 miles (900 km) of affected shoreline, "which is too large an area—too wide—for a meteorological origin." Add to this the fact that most typhoons hit Japan between August and October. These mystery waves had arrived in the dead of winter. "We examined the weather of that day and we found that the weather wasn't that bad," he said, with sunny or cloudy skies reported in most of central and northern Japan on January 27, 1700.

Having ruled out storms, they were convinced it had to be a tsunami and it must have come from across the Pacific. So they examined another catalog of historical temblors that listed the ones from South America. "What we found was there were earthquakes before and after 1700," remarked Satake. "For example, in 1687 there was an earthquake recorded in South America and a tsunami from this event was also recorded in Japan. And similarly in 1730, we have an earthquake in South America and following that a tsunami was also documented in Japan. But there's no earthquake in 1700. So we could rule out the possibility for South American origin."

Judging by the reported height of the waves in Japan, the orphan tsunami was too big to have come from Alaska, the Aleutians, or the Kamchatka Peninsula. The way the plate boundaries line up, those subduction zones are almost parallel to Japan rather than perpendicular to it. So movements along those faults generate waves with primary energy vectors that usually miss Japan. The 1964 event in Alaska, for example, sent large and destructive waves to Vancouver Island and California, but the side-angle waves that hit Japan were less than a foot (0.3 m) above the tide.

To make the much larger waves (some as high as sixteen feet, or 5 m) that ran up the beaches of Japan in 1700, a quake in Alaska or Kamchatka would have had to be magnitude 9 or larger. But there were no written records or geologic evidence for a megathrust event of that size in either region in January of that year. Therefore, by process of elimination, Satake and his colleagues concluded that Cascadia was "the most likely source" of the orphan tsunami.

To estimate the magnitude of the parent earthquake, Satake compared the damage reports from the orphan tsunami with those from the 1960 Chile wave, which killed 140 people in Japan, and found the wave heights to be similar. To create a tsunami as big as the one that came ashore in 1700, Satake and his team figured the Cascadia earthquake must have been at least magnitude 9 or larger. From orphan tsunami to parent quake, these geologic sleuths had traced the last big failure of Cascadia's fault. The next step was to figure out the exact time it happened.

They knew from other research how fast a tsunami could move. "The deep-ocean speed is approximately the same as a jetliner," Satake told me, "so it would have taken about nine or ten hours from the West Coast to Japan." Next they took the confirmed arrival times of waves hitting the beach in Japan and calculated *backward* to figure out when the earthquake must have triggered the tsunami.

"The earliest documented tsunami arrival time was around midnight on 27 January, Japan time," wrote Satake and his colleagues in *Nature*. "Because tsunami travel time from Cascadia to Japan is about 10 hours, the earthquake origin time is estimated at around 5:00 on 27 January GMT or 21:00 on 26 January local time in Cascadia. This time is consistent with Native American legends that an earthquake occurred on a winter night."

And so the conceptual or philosophical barrier was finally breached. It would no longer be possible to argue that Cascadia was probably harmless. The doubters could not point to an absence of great earthquakes in "all of recorded history" and make their case for aseismic subduction. Thanks to a handful of local officials in Japan, the written history of North America's west coast had been extended back more than three centuries—far enough to reveal a new and ominous picture of this hidden plate boundary.

The evidence that Cascadia's subduction zone had generated earthquakes big enough to send damaging waves all the way to Japan seemed quite convincing. All of Brian Atwater's tenacious digging in the mud, together with Kenji Satake's eureka moment and deductive reasoning, had paid off. "We were really very excited to find this evidence," Satake said with a modest smile.

"It used to be that we would say that Cascadia made an earthquake and tsunami, or a series of earthquakes and tsunamis, *about* three hundred years ago," Atwater began to tell coastal audiences at tsunami-safety workshops. "Now, we can say that at *9:00 p.m.* on the *26th of January in 1700* it made one giant earthquake and a trans-Pacific tsunami. That gives this history a more definite feel than it used to have."

Like diligent researchers everywhere, though, they knew the quest for confirmation was not over. Convincing and logical evidence they had. But was there any other way to explain the data? Could any part of their hypothesis be falsified?

Satake sent a draft of his paper to half a dozen other researchers for reviews. His fellow scientists made suggestions for fine-tuning but gave it a thumbs-up. The editors of *Nature* liked what they saw enough to publish the breakthrough finding on December 5, 1995. The next phase of the investigation would be a thorough search for the original documents written by samurai warriors, local merchants, and public officials in villages along Japan's east coast.

The story of how Brian Atwater and Kenji Satake found a way to work together brings a grin to my face. When filming the story of the orphan tsunami, I got caught up in the blind alleys of a detective story, fascinated by the logic and niggling details, constantly asking myself—how the heck did these guys figure this out? Along the way I forgot to ask how they tackled the research from such different scientific and cultural perspectives while separated by thousands of miles of North Pacific Ocean. Satake dashed off a quick email that gave me the context and the smile.

"As a traditional and sometimes stubborn geologist," Kenji wrote of Brian, "he would not be convinced until he examined the data by himself. But the 'data' in this case are Japanese historical documents, not coastal geologic sections [in which] he was specialized. A normal person would not step further. Brian, however, was different. He first attended a class at University of Washington to learn Japanese. Obviously, learning a new language at an age of nearly 50 is not an easy task, and Japanese is not an easy language for westerners . . . He spent about a year in Japan [1998–99] to study the historical documents and the social and historical backgrounds of the documents. Our goal was to write a book for the U.S. general audience to introduce the Japanese documents recording the Cascadia tsunami."

Satake, Atwater, and their colleagues reexamined original documents in the six main villages and towns that had recorded the tsunami's progress down Japan's eastern seaboard. The waters arrived late at night on the north coast and worked their way south as the sun rose. In Kuwagasaki, a seaport village with roughly three hundred houses in 1700 where Cascadia's wave first made Japanese landfall, the midnight flood and ensuing fire destroyed one-tenth of the homes. As villagers fled to higher ground the frigid water wrecked thirteen homes outright and set off a blaze that burned twenty more. The entry was written by a local magistrate, who noted the exact arrival time of the incoming tsunami.

At the south end of Miyako Bay the water washed away houses along the shore and then entered Tsugaruishi village just over half a mile (1 km) inland, where it caused panic among the residents and sparked fires that burned another twenty-one homes. An independent report written by a local merchant in his family's notebook mentioned that there was no ground shaking—which must have seemed odd since this damaging mound of water looked so much like a tsunami.

From Tsugaruishi the wave continued another half mile inland and across a floodplain to a place called Kubota Crossing, and on to the foot of a hill that has long been capped by a shrine to the Shinto god Inari. This put the high-water mark for the 1700 tsunami at roughly the same place as the train of waves that came from Chile in 1960. That means Cascadia's wave was about sixteen feet (5 m) high when it crashed to shore.

Farther down the coast at the river port of Nakaminato, around eight o'clock on the morning of January 28, high waves—probably strong outflow currents from the ebbing tsunami, churning against ordinary ocean swells at the river's mouth—held a cargo vessel offshore and prevented it from entering port. From nearby Miho came a report that "something like a very high tide" swept ashore seven times between dawn and ten o'clock that morning. At the end of day the vessel was still offshore when a storm blew up, broke the anchor lines, drove the boat against a rocky beach, and sank it.

A cargo of thirty tons of rice, bound for the capital in Edo, was destroyed and two crewmen were killed. In what amounts to a police report, the boat's captain and two villagers wrote descriptions of the sinking and petitioned a district magistrate to certify the circumstances of the accident that ruined 470 bales of rice thought to be owned by two samurai from the north coast. The official port certificate absolving the crew of responsibility for the Nakaminato shipwreck of January 28, 1700, was among the documents dug up and examined by Satake, Atwater, and their colleagues.

One village headman, writing his eyewitness account, described seven surges of water that moved "with the speed of a big river." He took the precaution of sending elders and children to the safety of higher ground at a local shrine and then consulted village elders about the puzzling lack of an earthquake.

Kenji Satake arranged to meet me in the seaport city of Tanabe, southernmost of the six communities where records of Cascadia's wave had been discovered. In 1997 Tanabe had a population of 72,000. From a distance the harbor and skyline, set against a range of low, blue-green hills and small mountains, looked to me a lot like the coast of southern British Columbia, where the orphan tsunami was born.

Tanabe was badly shaken in 1946 by the magnitude 8.1 subduction rupture along the Nankai Trough, less than sixty miles (100 km) away. The tsunami that hit the harbor only a few minutes later swept 145 homes away, damaged or destroyed 1,200 more, and killed 69 people. Smudgy black and white photos of the aftermath look similar to those from Phuket and Banda Aceh in 2004—jumbled piles of splintered lumber and debris, fishing boats tossed high and dry, stunned survivors looking lost, wandering in the rubble.

A group of city officials took us up a stone stairway leading to an ancient temple on high ground. Twenty steps up the mountain we saw a monument to the dead, engraved with a notch in the marble tablet to indicate the high-water line of the 1946 tsunami. I tried to picture a wall of water high enough to touch the eaves of a two-story building, because that's how far up it looked to me. Satake and his team compared the 1946 inundation zone to descriptions of the 1700 Cascadia waves and concluded that they had probably reached this height as well.

Tanabe's mayor in 1700, a member of the Tadokoro family, lived in the merchant district, north of and slightly uphill from the old walled castle. The city's population then was roughly 2,600 people. The mayor himself escaped the worst effects of the rising water because the flood stopped roughly five hundred feet (150 m) below his home. As an eyewitness, Tadokoro collected and supervised the writing of the official damage reports.

With brushes and black ink faded only slightly after three hundred years, he or his scribe told the scary story in fluid, artistic strokes. When we arrived to film the evidence, Japan's seismic historians had already searched thousands of pages of the city's official ledger as well as the Tadokoro family's "diary of ten thousand generations" to find the orphan tsunami. Complex ideograms flowed in neat, vertical columns on washi, a traditional acid-free paper strengthened with fibers of bark. Over the course of more than three centuries the smooth, durable pages had survived water damage and were perforated here and there by bookworms. But the essential details remained.

The rising Pacific surge flooded the castle moat and ruined nearby paddies and fields of wheat. It soaked and destroyed bales of rice stored in several government warehouses. In those days rice was almost the same as money. On behalf of the shogun, samurai collected taxes in the form of rice. Therefore it was very likely that the shogun's rice got wrecked by the orphan tsunami from Cascadia. The documents describing what happened to the shogun's rice are now considered proof that Cascadia is a dangerous subduction zone.

But there was still one more question to answer. "Kenji's story seemed to hang together very well," David Yamaguchi agreed, "but it was still very much a story that was hypothetical. He could not prove conclusively that that wave came from here," the Pacific Northwest coast of America.

To do that something a little more tangible was needed. And somewhere along the way another eureka moment happened. If radiocarbon would never be precise enough by itself, and if the outer wood of those trees in the ghost forest was too chewed up to give the exact year of death, what about digging deeper? Why not dig up the *roots* of those standing dead cedars and look for that final year's growth ring? Would the roots be preserved well enough underground to show a ring for 1699? If so, wouldn't that suggest the trees did not live long enough to generate a growth ring for 1700? That's the question Brian

Atwater asked himself, and he called on David Yamaguchi's expertise one more time.

"Brian asked me for leads," Yamaguchi recalled. "He asked me to choose my top-ten list of trees that I thought gave me the best tree-ring dates—that I was most confident in. And then he went out with a team—a young team—and excavated these trees down to roots where they found that they had intact bark."

At the ghost forest, Atwater packed his pickup truck full of muddy roots and a gang of exhausted diggers with aching muscles and drove home. Back in Seattle he sanded the samples in his backyard, checked them with a hand lens, and then packed them off to the Lamont-Dougherty lab in New York. "Trees only put on rings between about May and September," Yamaguchi explained, "and so if they were killed by an earthquake in January of 1700, their outermost rings should be 1699."

And that's exactly what they found. Yamaguchi matched the root rings against "the barcode" of rings from the witness trees and confirmed—beyond any reasonable doubt—that the cedars of the ghost forest were killed in the winter of 1700. "Radiocarbon dating can't do it. Chemical dating can't do it. It was up to tree rings," Yamaguchi proclaimed with a wide smile.

All that wet and dirty shovel work had finally paid off. "This finding here at this estuary," said Atwater, standing on the bank of the Copalis River, "and at the other three estuaries in Washington to the south of us—all having trees dying between August of 1699 and May of 1700—that kind of long reach of coast all dying together at the time of the solitary tsunami in Japan makes a very big earthquake at our Cascadia Subduction Zone more plausible than it was before."

For Atwater it was pretty clear from the beginning that a rupture had happened here and that a big tsunami had been generated. The evidence had always been there in the geology along the west coast. Cascadia's fault had indeed written its own history in rock and mud. "It just becomes reinforced by having a human record in Japan that says there's something corresponding over there—that people observed, and wrote about and that got preserved in those wonderful records."

Together Atwater, Yamaguchi, Satake, and their colleagues had carried the investigation to what seemed like a logical conclusion. They tracked down the parent earthquake of Japan's orphan wave of 1700. It had been an exciting, frustrating, and tedious journey that ended with deadly implications.

How did it feel to learn something so dire? "It's mixed feelings," admitted Yamaguchi. "As a scientist, it *was* sort of a eureka feeling— *wow*, we've worked this out! As a resident of Seattle, it's also sort of a humbling and scary kind of feeling because you realize this is historical evidence for an earthquake of a size this region—and the world—has scarcely seen."

"Once we admit this earthquake happened in the past," Satake told me, "we also have to accept that the same earthquake will happen in the future. Because earthquakes in subduction zones—we know that they repeat. So the most significant thing—the most important thing that comes out of this research—is that now we know that this earthquake can happen in the future."

CHAPTER 18

Episodic Tremor and Slip: Tracking Cascadia with GPS

At four o'clock in the morning on August 31, 1983, the ramp crew at Anchorage International Airport finished refueling a Boeing 747 operated by Korean Air Lines. A few minutes later KAL flight 007 took off on the final leg of its journey from New York to Seoul. As the sun rose the big jet roared across the International Date Line and headed south and west toward its home port. Then, for reasons never verified, the plane allegedly strayed off course over the Sea of Japan west of Sakhalin Island and was shot down by fighter jets from the Soviet Union.

All 269 passengers and crew aboard KAL 007 were killed and for several weeks the Cold War threatened to escalate into something worse. When flight data recorders were recovered from the wreckage, the Soviets refused to release them to international aviation authorities investigating the crash, so it was impossible to prove or disprove the conflicting stories of why the Korean jet had drifted into prohibited airspace. But after the heat of the moment faded, one positive change did come about as a result of the tragedy.

To reduce the odds that a navigational error could ever cause something like this again, President Ronald Reagan ordered the U.S. military to make its NAVSTAR Global Positioning System (GPS) available for civilian use. GPS did not change the world of aviation overnight because there still weren't enough satellites in orbit in 1983 to make the system fully functional. The precision of the new and still-evolving technology would eventually have a significant effect on everything from tracking animal herds in the wilderness to finding a freeway exit to the nearest pizza parlor—and measuring the drift of continents.

By figuring its distance from at least four of the NAVSTAR satellites, a civilian-made GPS receiver could calculate its position anywhere on the surface of the earth to within roughly a hundred feet (30 m). The system was based on the same principles of trilateration that Jim Savage and Herb Dragert and others had used to measure the shift of mountain peaks in Puget Sound and on Vancouver Island. Their Geodolites had calculated the distance between two brass survey markers by measuring how long it took a laser beam to travel from one peak to another, bounce off a set of reflectors, and return to the starting point. GPS positioning did pretty much the same thing.

By 1992, when Mike Schmidt of the Geological Survey of Canada climbed Mount Logan with one of the slightly clunky but then state-of-the-art units, he was able to measure latitude and longitude with an accuracy of a few millimeters. They could nail the altitude within about ten or fifteen millimeters (0.4–0.6 inches). As Schmidt and Dragert and their geodesy team from the Pacific Geoscience Centre built their array of permanent tracking stations to measure the movement of Vancouver Island, enough new GPS satellites were being launched to make measurements far more accurate than laser beams bounced between mountain tops had been. The enhanced precision would also create one of the most baffling new mysteries Cascadia watchers had ever seen—a mystery once solved that would stun even those who thought they knew what to expect from an active subduction zone.

∞

"We have now established that this site," said Dragert, pointing at the Albert Head GPS monument (near Victoria on Vancouver Island) behind him, "that particular point right there, is moving at a rate of about six millimeters per year towards Penticton, roughly in a sixty-degree azimuth direction."

While six millimeters of horizontal creep didn't sound like much—less than a quarter of an inch—a dose of context changed the picture. Given that the last major earthquake to relieve stress on Cascadia's fault happened more than three hundred years ago, and given that stress has been rebuilding ever since, that tiny increment of eastward squeezing had been accumulating in the rocks of Vancouver Island and all along the west coast for three centuries. Dragert did the math for us. "The six millimeters times three hundred years is 1,800 millimeters. So 1.8 meters [6 feet] will be released at this particular location during the course of the next large earthquake."

In other words, if Cascadia had ruptured right then and there—after three hundred years of stress accumulation—the entire city of Victoria would have rattled, rumbled, and slid sideways nearly six feet (1.8 m) back toward the west as the strain came out of the rocks. The longer it takes the zone to fail, the more stress there will be—ready to snap in the next great rupture. "We proved that the margin was deforming," said Dragert emphatically. "The mountains were indeed being squeezed landward. So it's not a hypothetical problem; it's a real problem. We are gaining strain energy all to be ultimately released by a large megathrust earthquake."

The switch from lasers to GPS technology made it possible to track tectonic movement 24/7 without having to wait for budget approval to hire another helicopter. "You didn't have to fly to the mountaintop—you just set up the instrument," Dragert enthused, "a totally automated instrument that told you what its position was day after day after day. It worked under any weather conditions. Much cheaper. I mean we basically had to give up [laser] trilateration because it turned out to be too expensive," he explained. "With this new technology we could measure even longer baseline distances—not just fifty kilometers, hundreds of kilometers—to a precision of one or two millimeters."

The Pentagon's new technology gave scientists a close-up view of tectonic motion that had been inconceivable thirty years earlier, when the debate about continental drift began in earnest. GPS proved not only that plates were shifting but how fast, in which directions, and how high the outer coastlines were rising, bending, and buckling. Before another year had passed, the satellites would also reveal to Dragert and his colleagues the next big secret of subduction.

"Without GPS technology we would never, never have observed 'silent slip,'" he said, with fingers putting quotation marks around the term. The first time he spotted a tiny "backward movement" of Vancouver Island, it was so subtle, so apparently insignificant, it just had to be a glitch. At least that's what Dragert thought at first.

He remembered the bafflement as if it were yesterday. On a desktop computer screen he opened a file that showed the data from the station nearest Victoria. "We were kind of looking back over the last four years of data in '96. And we said, 'Gee, there's this funny offset,' right around October '94. So, what's going on?" His finger traced the steady, almost straight-line movement of the GPS antenna at Albert Head.

As expected, it had been creeping relentlessly east day by day for months, when all of a sudden there was a sawtooth zigzag in the data. The Albert Head antenna had apparently *switched direction* and doubled back in the opposite direction for about ten days. A concrete tower built into the solid bedrock of Vancouver Island had inexplicably stopped moving toward Penticton and was temporarily sliding back *west*. After the ten days, it started moving east again.

Because only one of the four existing stations had displayed this maverick behavior, Dragert convinced himself the data had to be wrong. "So I said, 'Okay, most likely our monument is unstable.' For some reason, even though this is in concrete—we have rebar drilled into the bedrock, so there's a very good coupling to the bedrock, and it's very competent bedrock, it's not fractured or weathered—so, strange as it may seem, maybe we just didn't see something. Maybe there's a fracture zone somewhere that we were unaware of and it's tilted our monument."

To find out what had gone wrong, Dragert and his colleagues drove back out to Albert Head, set up a laser transit, and resurveyed the antenna. "We did exactly the same survey as we did in '92 [when the system was first installed]. And according to both surveys—'Hey, the monument hasn't moved at all!' Less than 0.3 millimeters was the difference between the '92 survey and the '96 survey." The GPS tower itself was locked solidly in place, so it had to be the *ground* that was moving back and forth. In essence, Vancouver Island was being shoved to the east most of the time, but every fourteen months or so it would slip backward as if the underground stress had somehow been temporarily released or reversed.

How could that happen? Dragert laughed heartily. "We couldn't explain it. We simply said, 'That's life,' and we went on."

But Dragert and his colleagues kept watching the incoming data, determined to solve the mystery. Then one day—there it was again. The same apparent glitch, the same displacement. But now they had fourteen continuously monitored GPS stations in the Pacific Northwest, giving them much more precise data.

So Dragert started calling around to find out if any of the other research teams had seen anything this weird. Sure enough, when asked to take a closer look at their data plots, several of them saw a similar kind of reversed movement. Six other GPS antennas in both Canada and the United States had "jumped backward." In all a cluster of seven adjacent sites strung out across southern Vancouver Island from Ucluelet to Nanaimo and to Victoria and down Puget Sound as far south as Seattle had suddenly slipped backward in what looked like a slow, silent earthquake that took anywhere from six to fifteen days to happen.

And it was definitely a geographic cluster rather than a random scatter. The antenna up at Holberg, on the north end of the island, didn't move at all. Neither did the towers at Williams Lake or Chilliwack, British Columbia, or Linden, Washington. Only the stations in the middle moved. The stations at the extreme north and south ends of the GPS array didn't flinch.

Not only that but the slippage had started a few days earlier down in Washington State and then moved gradually toward Vancouver Island—almost like the slow-motion unzipping of a fault. Dragert beamed. "We were saying, 'Holy crap, this is great! This is absolutely great!' This not only told us that the signal was real, it told us the signal was constrained to a given area. And it took time to travel from the south to the north." Basically the signal looked like a ripple moving through rock.

"It was like something migrating underneath our feet," he said. There was still no logical explanation, however, for the odd, backward-jumping movement of a handful of GPS antennas, so a professional skeptic's first response was to say it still was probably some kind of mistake. If the GPS monuments are locked in solid rock, then something else must be wrong. Before publishing their data, Dragert and his colleague Kelin Wang had to rule out every conceivable analytical glitch and recalculate all the GPS orbits, just to be absolutely sure what they were seeing was real. Eventually they arrived at the conclusion that the silent backward slip was not a fantasy.

On his computer screen Dragert again traced the upward-slanting line of data with his finger. His hunch was that somehow, way down in the lower part of the subduction zone, a small measure of tectonic stress was being temporarily relieved. The deepest part of the zone had—in his words—come unsprung. It had slipped.

When he expanded the timeline to display several years of continuous movement, the zigzag pattern became even more obvious. The reversals put spikes in a straight-line graph that made it look like a saw blade with evenly spaced, sharp teeth. The thing that struck me about it was the regularity. How could anything in nature be that punctual? Why would it keep coming back every fourteen months?

"Interesting question," Dragert replied. "We have no idea."

In a paper published in *Science* on May 25, 2001, Dragert and Wang released their mysterious findings to the world. "In the summer of 1999, a cluster of seven sites briefly reversed their direction of motion," they wrote. "No seismicity was associated with this event." Meaning there were none of the normal seismic shockwaves one would expect from something that looked in every other respect like an earthquake.

Could two plates really slip *without* producing detectable seismic waves? Dragert and Wang noted that all the GPS tracking stations that moved backward were some distance to landward of the locked part of the zone. So whatever was causing the backward movement had to be happening way down deep, where the rocks were hotter, softer, and less likely to stick together for long periods. They calculated that if an area 30 by 190 miles (50 by 300 km) were to move backward less than an inch (2 cm), the fault would still be generating the energy equivalent of a magnitude 6.7 earthquake. But where was that energy going?

They concluded that these slip events were probably transferring stress "uphill" to the shallower part of the locked zone in "discrete pulses." So even though nobody could feel them at the surface, each time one of these bizarre reversals happened, it was probably pushing the fault one notch closer to failure—a giant earthquake.

The mystery of "silent slip" took another unexpected turn when Herb Dragert traveled to a science conference in New Zealand, where he learned from Kazushige Obara that something very similar was happening in Japan. "He's the one that discovered tremors," said Dragert, "but he didn't know what they were. He had no idea there was crustal displacement—crustal motion—involved with these. He just kind of said, 'Hey, these are weird signals that aren't earthquakes, but they're not volcanic.' So he called them deep, non-volcanic seismic tremors."

After quizzing Obara about the details, Dragert returned to the Pacific Geoscience Centre and started hunting for a connection between *tremor* and *slip*. Garry Rogers, by now one of Canada's top seismologists, had an office just upstairs and down the hall. Dragert provided Rogers with dates when the zigzag patterns had showed up on the GPS. Rogers then dug out boxes of seismograph records from the PGC archives and both were stunned to find a near-perfect match-up.

"I opened the box," said Rogers, recalling the search, "and *there* was the tremor!" He showed me the seismogram and pointed to a squiggle of tremor noise that coincided with one of the backward jumps on the GPS. "Herb gave me the next date—I opened the next box—and *there* was the tremor event. And boy, the hairs just stood up on the backs of our necks."

"Oh yeah," Dragert said, beaming. "Yeah, that was exciting. Every time a slip event occurred, there was a huge increase in this background noise, a huge increase. And so it was at that point the eureka came through. We said, 'Hey, these things are intimately related.' We knew we had something."

Condensing it all into a neat little sequence, Rogers explained that the deepest part of the fault—way down where it's hot and gooey— fails, or slips loose, every fourteen months. When it slips—for a period of about ten days—the GPS antennas on the surface record a backward jump. The land actually recoils as the fault slips, and the seismographs record a silent tremor. Fourteen months' worth of deep tectonic stress is transferred upward into the colder, harder rocks of the part of the fault that has remained stuck. Then, with the stress transferred, the lower part of the fault locks up again and the cycle repeats itself. "It's a very unique phenomenon," Rogers said. "We called it episodic tremor and slip, or ETS for short."

The most fascinating revelation in all of this was that the upper part of the fault—the part that's been locked in place and building stress ever since the last great Cascadia earthquake more than three hundred years ago—gets another increment of stress added to its load on an incredibly and mysteriously regular basis. Almost like clockwork. In other words, the stress doesn't just add up gradually until the fault ruptures; it comes in discrete little jolts every fourteen months.

"It's like adding straws to a camel's back," Rogers suggested. "Probably when one of these straws are added it will break the camel's back." And we'll have the megathrust earthquake we've all been waiting for. But maybe—just maybe—with this bizarrely regular timing of the ETS events, there might be a way to anticipate that quake.

CHAPTER 19

Turbidite Timeline: Cascadia's Long and Violent History

"Initially, my colleague Hans Nelson and I didn't believe it," said Chris Goldfinger with a smile. "We thought it was probably wrong. It was way too simple. It can't be right. So we wrote a proposal to the NSF [National Science Foundation] to go prove John Adams wrong."

This was Goldfinger holding court in the ship's lounge off the coast of Thailand en route to Sumatra to collect piston cores of ocean mud from landslides triggered by the catastrophic quake and tsunami of Boxing Day 2004. He was setting the scene for those who had not yet experienced the intrigue and frustration of trying to study deadly subduction zones that lie hidden beneath the sea. For Goldfinger and his research partner at Oregon State University, Hans Nelson, the story had begun back in 1985 when John Adams, at the Geological Survey of Canada, wrote a controversial paper based on some old OSU core samples.

The original work in question had been done in 1968 by graduate student Gary Griggs and his thesis advisor, LaVerne Kulm, who (along with Bob Yeats) later became Goldfinger's thesis advisor as well. The Griggs and Kulm piston cores revealed a series of undersea landslides that had traveled hundreds of miles downhill from the edge of the continental shelf into deeper ocean water along a network of seafloor canyons and channels off the coasts of Washington, Oregon, and northern California. The question in 1970, when the original data were published, was what had caused the landslides.

The core samples had been gouged with steel tubes and plastic pipes from a web of deep-sea channels many miles apart, each showing evidence of thirteen or more landslides. Griggs and Kulm over beers after work one night came up with several possibilities. Either the sediment flows "self-triggered" once the seafloor mud had piled up deep enough to collapse under its own weight. Or big storms with very deep waves might have done it. Or perhaps it was big earthquakes.

Griggs and Kulm knew from looking at the core samples and measuring how thick the layers were and how far they were from the Mazama ash layer that each of the thirteen slides had happened within minutes of each other—all up and down the coast. How could so many turbid flows happen in so many canyons so far apart all at once? If it were mere coincidence, the same coincidence had happened thirteen times, once every six hundred years. In one of a series of papers on the subject, Chris Goldfinger would later refer to this as "coincidence beyond credibility."

The self-triggering idea seemed the least plausible because it was unlikely that sediment would accumulate at exactly the same rate along hundreds of miles of the continental shelf. Rivers of different size and volume dump differing amounts of debris in different places along the coast at different rates—so how would piles of sand so far apart all collapse into turbidity currents at exactly the same moment? Thirteen times?

Storms with waves big enough to trigger deep-sea landslides also seemed a tad improbable, especially on such a wide geographic scale. They were pretty sure that turbulence from waves generated by winter's worst storms generally did not reach that far below the surface, so how could they disturb heaps of sand at the heads of steep canyons (where most of these landslides began) that were anywhere from 500 to 1,300 feet (150–400 m) deep? If storm-triggered landslides had happened, the odds were higher that they would have occurred at different times in different places—not all at once, as the core samples showed.

Seismic shockwaves also seemed unlikely culprits because in 1970 there was no historical record of megathrust ruptures in the Pacific Northwest. When one of the grad students casually suggested the quake hypothesis that night over beers at OSU, Professor Kulm legendarily replied, "Nobody would believe it." Thus the earthquake story entered a state of limbo, with no obvious way to prove or disprove it.

Goldfinger can still recall the doubt and disbelief that flashed through the corridors of the marine geology department when the first John Adams paper appeared in the scientific literature saying, in essence, that Griggs and Kulm should have ignored the potential skeptics. Without going to sea, without collecting a single new sample himself, Adams, the outsider, this transplanted New Zealander who had moved to Canada via Cornell University, wrote a paper based on the Griggs and Kulm cores and concluded that seismic jolts were the best—indeed the only logical—explanation for the thirteen turbidites.

It must have seemed as though a foreign spy had raided the OSU lab and stolen the thunder of the home team's original discovery. As it turned out, not a bit of skulduggery was involved. Adams simply wrote to OSU officials asking permission to look at some of the core logs that had been sitting in storage since 1968. Picking up the storyline more than three decades later, Chris Goldfinger told his colleagues aboard the *Roger Revelle* off the coast of Sumatra how an amazing feat of deductive reasoning had come about.

"He saw the same thing [that Griggs and Kulm had seen], that there were thirteen turbidites above the Mazama ash. So he tried to come up with a method to prove or test the origin of these things. And what he did was this: he noticed that there's a confluence of two channel systems right through here."

Goldfinger pointed to the map beamed from his computer via an overhead projector to a screen at the front of the ship's lounge. "Well, it turns out that all of these cores have thirteen turbidites." He pointed to sampling sites on both main channel systems. "So here's his little test, right here. He said, 'Well, okay—if you have two channels or two canyon systems, 200 kilometers [125 miles] apart, and one has thirteen turbidites and the other has thirteen turbidites—how could you possibly pass the confluence and go downstream and *not* have *twenty-six* turbidites?"

Goldfinger scanned the faces around the room to see fascinated smiles and raised eyebrows. "You *should* have twenty-six turbidites here, right?" Landslides send thirteen turbid currents of sand sloshing down two channels that run together, so there ought to be a total of twenty-six turbidite layers in the mud downstream from the confluence.

"And there are only two ways you could *not* have twenty-six," he continued. "One is that it just coincidentally dropped out thirteen of them [for unknown reasons, thirteen of the landslides didn't make it past the confluence], which didn't seem very likely. And the other way is that they arrived at the confluence *at exactly the same time,* plus or minus about five minutes—and merged." In other words, if a big quake triggered a landslide at the same moment at the head of each of the major canyons along hundreds of miles of the continental shelf, then all the mud flows would cascade downhill synchronously and would arrive at the downstream confluence where all the offshore sea channels meet— at the same moment.

Below the confluence there would be a single, merged turbidity flow. No matter how many small tributaries fed into the main channel from the steep slopes above, the total number of turbidites *below* the confluence would always be the same: one for each coastwide landslide. The only reasonable explanation for so many landslides happening synchronously along so many miles of coastline was large subduction earthquakes.

"And that's what John argued," said Goldfinger. And then he paid one of the highest compliments one scientist can offer another. "This was done purely with thinking power. And that's out of fashion these days." The praise seems all the more significant given that Goldfinger and Nelson had initially doubted the simplicity and elegance of the Adams hypothesis enough to ask the National Science Foundation for research money to prove him wrong.

Goldfinger had gone to sea in 1992 and discovered the under-water Elvis, a man-with-a-guitar–shaped mound of folded and fractured ocean sediment, along with eight other new faults in the upper plate along the continental shelf. He wrote that the rough edges of these fractures may limit the size of Cascadia's big subduction quakes by inhibiting the build-up of strain energy and concluded that Cascadia may be the type of subduction zone at which magnitude 9 events "do not occur." Without magnitude 9 ruptures, the Adams hypothesis had to be wrong. Smaller jolts just wouldn't do what the hypothesis demanded.

But a subsequent research voyage in 1999 turned things around for Goldfinger and Nelson. The evidence in favor of big landslides was very obvious in the offshore mud when examined close up. Not only did they confirm the same thirteen turbidites along 375 miles (600 km) of coastline, but they ventured farther north to the Nootka fault, at the upper end of the Juan de Fuca plate off Vancouver Island, and farther south all the way down to Cape Mendocino in California. Along the way they collected nearly a hundred new cores and added several discoveries of their own, extending the count from thirteen to eighteen events—presumably large quakes—and extending the timeline back to the end of the last ice age, roughly ten thousand years ago. They saw these dark, sandy landslide scars on the ocean floor as "earthquake proxies," the telltale markers of Cascadia's long and violent past.

In the lounge aboard the *Roger Revelle*, Goldfinger explained the challenge of adding the five new turbidites to the series of thirteen already established. The problem with the new samples was that they came from offshore river channels that were not physically connected to the network of channels flowing primarily from the Columbia River and the Strait of Juan de Fuca up the coast. The new evidence was found in a completely separate, unrelated grid of outflow channels from Barclay Canyon, off Vancouver Island, and from the Rogue River Canyon, midway down the Oregon coast.

How would it be possible to know whether the five additional turbidite flows had happened all the way down the coast at roughly the same time, in the same kind of synchronous gushes that Adams noticed in the main Cascadia channel? The question could be answered using oilfield techniques well known to many of the faculty at OSU, where petroleum geology was a significant part of the academic program. As Chris Goldfinger likes to tell it, oil drillers have been doing this sort of thing for years.

Again he pointed to the overhead map, zeroing in on the offshore region near the Oregon–California border. "These channel systems don't have the same sources and they're even further apart [than the channels that Adams studied]. They don't have anything in common," he said. Oregon's Rogue River, for example, flows directly from Crater Lake—the former volcano Mount Mazama—to the sea with no downstream connection to the Cascadia channel. There is no confluence of canyon heads and tributaries that would physically link the Rogue turbidites with the others farther north.

The *stratigraphic patterns* in all the samples, however, did look very nearly identical. The relative age, thickness, and spacing of the alternating bands of turbidite sand, silt, and gray-green ocean mud were the fingerprints of Cascadia's history. Goldfinger and Nelson used a process known as wiggle-matching to make a detailed, layer-by-layer examination of all the minute gradations of muck that had been laid down on the ocean floor.

"Correlating the wiggles" in core samples from the entire length of the Cascadia Subduction Zone took quite a while, but the match-up was pretty convincing. "Even though this hadn't been used before in paleoseismology," Goldfinger said, "this is basic, subsurface oilfield geology. This is how oil deposits are tracked from one place to another, because turbidites make good oil reservoirs of sand. So correlating tur-bidites from place to place is something that hundreds of people do on a daily basis all over the world. We're just taking that technique and applying it to a different purpose here." Instead of chasing oil, they were chasing earthquakes.

By 2003, when Goldfinger and Nelson published another paper based on more turbidite data, the tide of opinion had turned. The number of doubters had dwindled. The onshore record of sunken marshes, drowned trees, and sheets of tsunami sand had been accepted by most as evidence of Cascadia's past quakes. The evidence from offshore tur-bidites was still circumstantial, although the case was strengthened now that the work of John Adams had been redocumented, confirmed, and extended. Still, there was a lot to be done.

Goldfinger knew there were not enough data yet to establish abso-lute numerical ages for each of the offshore events—progress was slow because radiocarbon dating was difficult to do with so little plankton or other biotic material available in the deep-sea samples—so it was initially hard to correlate the turbidite record with the land-based data. There were enough similarities in the offshore core patterns, however, to establish "lateral continuity" of the turbidite layers. Meaning tur-bidite bed number three from a core taken off Vancouver Island was probably in the same regionwide stratigraphic layer as turbidite bed number three from a core taken near the California border.

Whatever triggered the offshore landslide up north presumably also triggered simultaneous landslides hundreds of miles to the south. The exact date may not be known, but in all likelihood the match-ing turbidites made a synchronized plunge downhill. And the only force strong enough to rattle the sea floor all the way from Vancouver

Island to California would have to be a very large temblor.

As a control sample, to see what the ocean mud looked like beyond the end of Cascadia's fault, Goldfinger and Nelson took their 1999 cruise ninety miles (150 km) south of Cape Mendocino, where they collected three more cores in the Noyo channel, an offshore canyon that drains the northern California continental margin adjacent to the San Andreas fault. They discovered a similar cyclical pattern of sandy turbidite flows interlaced with ocean mud, the main difference being that here the landslides seemed to happen more often. In the last ten thousand years there appeared to be thirty-one events—most likely caused by San Andreas ruptures big enough to trigger offshore landslides. It seemed that California's most famous fault was causing the same kinds of offshore landslides as the Cascadia Subduction Zone.

With ninety-six new cores going back nearly ten thousand years, there was finally a long enough history in the mud to look for patterns in the timing of these monster quakes. Suddenly, with the offshore landslide samples, the clock could be rolled back much further than before. At least in theory the new turbidite timelines offered a glimpse of long-term fault behavior. And the Cascadia cores did seem to reveal a repeating cycle. Judging by the thickness of ocean mud laid down between the turbidite layers, and with increasingly precise radiocarbon dates, they could tell roughly how much time had elapsed between events.

The first cycle began with a long quiet period after the Mazama volcanic eruption—more than a thousand years *without* an earthquake. After the first rupture was another moderately long interval of quiet, followed by two more jolts at shorter intervals—the shorter being only 215 years. The recurrence interval, or gap, between jolts was long, short, short. And that same sequence—long, short, short—had apparently repeated three times in the last 7,500 years.

Goldfinger and Nelson wrote that "while it is tempting to expound about earthquake clustering," it was still early days. They had taken "a tantalizing look at what may be the long-term behavior of a major fault system" but were careful to point out that their analysis was preliminary and would require confirmation of the radiocarbon age data.

Essentially they had encountered the same problem with radiocarbon dating that Brian Atwater did: there was very little biotic material to work with in deep-sea mud, and it had a way of getting moved around by burrowing sea creatures and sloshed out of the tops of the piston cores as they were wrangled onto ship's decks in heaving ocean swells. Unlike Atwater they could not rely on a ghost forest of ancient cedars conveniently nearby; nor could they use tree rings cored from perfectly preserved roots to help nail down the turbidite dates by other means.

They did, however, feel confident the wiggle-matching technique borrowed from their oilfield colleagues would eventually overcome these problems to help establish a solid and convincing long-term history for Cascadia. By December 2004 the clustering story had become more refined. Their turbidite studies had been updated, peer reviewed, and republished in several different science journals with enough new details that the media relations department at Oregon State decided to issue a news release. A draft prepared just before Christmas was set aside to be polished and sent out after the holidays.

Then, when no one was looking, another subduction zone in the Ring of Fire ripped apart, and the entire planet got knocked for a loop by a temblor so big it made the earth wobble slightly on its axis. The great Sumatra quake and tsunami of 2004 happened the day after Christmas and all eyes shifted instantly to the Indian Ocean. So it's unlikely that any more than a few diligent local reporters paid much attention to the OSU release issued on New Year's Eve. For those who bothered to read it, the release provided the latest chronology of clustered quakes on Cascadia's fault, a historical record with "two distinct implications—one that's good, the other not."

∞

Looking at the expanded pattern of turbidite beds off the coast of the Pacific Northwest, Goldfinger and Nelson concluded that the Cascadia Subduction Zone had experienced "a cluster" of four massive ruptures during the past 1,600 years. If the trend continued—if the pattern repeated—"this cluster could be over and the zone may already have entered a long quiet period of 500 to 1,000 years, which appears to be common following a cluster of earthquake events," noted the OSU release. That was the good news.

The alternative scenario was that the current cluster might still have one or more jolts left in it. The release pointed out that some clusters had up to five events and that within a cluster the average interval between earthquakes was three hundred years. By now most aware residents of the Pacific Northwest had heard the story that Cascadia's last megathrust shockwave was January 26, 1700, and therefore the next event might well be imminent.

"The Cascadia Subduction Zone has the longest recorded data about its earthquakes of any major fault in the world," wrote Goldfinger in the OSU release, putting his new ten-thousand-year turbidite timeline front and center. "So we know quite a bit about the periodicity of the fault zone and what to expect. But the key point we don't know is whether the current cluster of earthquake activity is over yet."

By December 30, 2004, when the OSU release went out to the media, the total number of Cascadia earthquakes identified in the turbidite cores had advanced from eighteen to twenty-one, at least seventeen of which had ruptured the entire length of the plate boundary from Vancouver Island to Cape Mendocino, causing magnitude 9 shockwaves and major tsunamis almost identical to the frightful scenario the whole world was watching hour after hour as the Boxing Day tragedy played out across the Indian Ocean. What we were seeing on television had happened here too—at least seventeen times.

When Goldfinger went to sea off the coast of Sumatra in the summer of 2007, the Cascadia turbidite data had been updated and refined yet again. The temblor count had gone up again as well. When I met him to film an interview for a documentary (called *ShockWave*) comparing Sumatra with Cascadia, Goldfinger revealed an all new and even more ominous summary of North America's biggest and potentially most violent fault.

"Right now, using all of the land evidence and all of the marine evidence that we have, we can put together a story going back about ten thousand years that shows a total of thirty-four earthquakes, plus or minus one or two," Goldfinger intoned somberly. The clustering story was there in the data but still difficult to verify. "It's something that's totally dependent on the radiocarbon ages and those have a lot of slop in them," he continued. "So it appears that events *seem* to cluster in time and then sort of die back and are quiet for a time—and then cluster again. I would say there's probably a 70 percent chance that there's some statistically meaningful clusters in the earthquake record."

There was no obvious explanation for why quakes might cluster. "We're just starting to really investigate the relationship between the Cascadia Subduction Zone and the faults that surround it," Goldfinger said. "It's of course connected to the entire Ring of Fire by other faults. We have the Queen Charlotte fault going off into Canada and the San Andreas fault going off into California, and all of these faults are all physically connected. So when you move one, it affects the others."

I had read a study by Ross Stein of the USGS in Menlo Park that showed how one earthquake could trigger another. The magnitude 7.5 Landers quake in southern California in 1992, fifty miles (80 km) north of Palm Springs, had triggered swarms of smaller jolts—at least sixty thousand aftershocks—as far away as Mount Shasta at the northern end of the state and even in Yellowstone National Park, more than eight hundred miles (1,300 km) away. Goldfinger told me the same thing was probably happening here in the Pacific Northwest.

"Even if they don't *trigger* each other, Cascadia and the San Andreas have stress relationships so that when one moves, it'll affect the stress state of the other. And so as the entire Pacific plate moves—and the whole Ring of Fire has its ring of subduction zones—everything that happens in one affects the whole system," he explained. "So hypothetically the Cascadia Subduction Zone, when it has its large event—it slips something like twenty meters [65 feet] or so—the energy from that is transferred to other places. And one of the places it could be transferred to is the northern San Andreas."

The thought of earthquake dominoes was hard to avoid. "It looks like somewhere around 80 to 90 percent of the San Andreas events have a Cascadia event associated with them," he said. "Almost every event on the San Andreas has come fairly closely in time—within forty years or so—of a Cascadia Subduction Zone event. So we could be looking at a scenario where Cascadia goes off, and then some relatively short time later the San Andreas goes off as well. And if that relationship is true, that has some implications for planning for the future."

In the summer of 2010, an even more comprehensive package of data from Cascadia boosted the earthquake count yet again. Now it's forty-one events in total, nineteen of which have been full-margin ruptures, and the average length of time between megathrust disasters has been recalculated at roughly every 300 years.

It sounded to me as though Goldfinger's turbidite timeline—perhaps combined with those ETS events that come every fourteen months—could eventually help emergency planners by shedding new light on the dark art of earthquake prediction. I would soon learn, however, that there are those who vehemently believe any time or money spent trying to *predict* a disaster is a complete waste.

CHAPTER 20

When's This Going to Happen? The Problems with Prediction

Boiled down to its essence, the argument against earthquake prediction is that nature is simply too complex and chaotic. There are too many variables, too many things happening deep under ground where we cannot easily see what's going on. No matter how closely we study the problem, no matter how many instruments we deploy, we'll never be able to anticipate a specific earthquake in a specific place at a specific time. Critics of quake prediction believe the considerable sums of money spent trying to warn people about something that's going to happen anyway would be better spent steeling ourselves for the shock. Put the money into mitigation instead. Build stronger buildings, dams, and bridges. Reinforce schools and hospitals to make sure they won't fall down.

The emphasis on mitigation certainly sounds logical and sensible, but scientists and the politicians who hand out research funding also know or suspect that a vast number of ordinary people really do want and expect to be warned. Many assume that predicting the next quake is—or should be—the primary goal and responsibility of all researchers involved in earthquake studies. Impossible as the dream of prediction may sound to skeptics, optimists tend to believe that if enough smart people are given the right tools to do the job, there's no problem science cannot solve eventually.

With Chris Goldfinger's newer, longer timeline, showing at least forty-one Cascadia quakes in the past ten thousand years, one might think enough data have finally been gathered to show a pattern. If there is a pattern, then a forecast of some kind should be possible. No matter how much scorn the doubters may have heaped, no matter how many false alarms have been sounded over the years, the optimists refuse to quit. There's always the promise of better technology, more data, faster computers that should be able to spot patterns in nature's chaos. Plus there's the tantalizing story of one earthquake prediction that was absolutely right.

Kelin Wang, a scientist working for the Geological Survey of Canada who spends most of his research time today working on Cascadia's fault, told me the story of the Haicheng earthquake in China in February 1975. It was prediction's golden moment. Wang had heard the heroic stories of Haicheng as a young student. In 2004 he returned temporarily from Canada to the land of his birth and took advantage of a new openness in China to dig up the details of what happened in the months and days leading up to the now famous prediction.

He and three colleagues interviewed many of the people involved in China's official quake prediction program and were given free access to a treasure trove of declassified documents, including thousands of pages of handwritten notes and data logs. The story they found is laced with political intrigue and, for skeptics, a lingering ambiguity. What follows is a condensation of what happened.

At 8:15 a.m., February 4, 1975, Cao Xianqing walked in to a hurriedly convened meeting of his local Party committee in Yingkou County, about three hundred miles (500 km) northeast of Beijing, and announced that "a large earthquake may occur today, during the day or in the evening." He urged county officials to "please take measures" to make sure people were evacuated from their homes and workplaces as quickly as possible. Although higher ranking authorities in the Liaoning provincial government would later take credit for orchestrating the prediction, most citizens of Yingkou and Haicheng Counties were aware that Cao—known locally as Mr. Earthquake—was the man behind the crucial first warning on that fateful day.

The warning itself and the subsequent saving of thousands of lives would become the stuff of seismic legend and folklore: the world's first and to date only successful earthquake prediction followed by an evacuation. Thousands of lives were saved. Researchers around the world heard reports—both censored and at the same time dramatized—of what happened that day in China. Everybody who felt the least bit optimistic about the ability of science to predict fault failures was extremely keen to learn more. Several foreign delegations were dispatched to study China's success.

A quick seven months later the National Academy of Sciences in the United States would publish a massive study entitled *Earthquake Prediction and Public Policy* suggesting that forecasting should be the "highest priority" because it could clearly save lives. The panel of experts who wrote the report took strong issue with the politicians and the few scientists who thought predictions and warnings might cause panic and economic disruption resulting in more harm than the temblors themselves. The Russians were already quite advanced in their own prediction studies, and the Japanese had launched the first phase of a large-scale prediction research program in 1965. China's reported success story injected a surge of scientific adrenalin into the hearts and minds of those who saw prediction as the holy grail of the new seismology.

A vocal few doubted China's claims, coming as they did at the height of the political turmoil of the Cultural Revolution. Skeptics assumed the claims were at the very least embellished, if not complete hokum based on voodoo science and luck. Chinese experts were slow to publish their findings in the open literature in part because "leaking secrets" to foreigners was a criminal offense, so there was precious little in the way of written documentation for the rest of the world to study. Until things changed in China, outsiders who believed in prediction would simply have to take the Yingkou–Haicheng story on faith.

Cao was a young carpenter who learned to read and write only after he joined the People's Liberation Army in 1947. He fought in China's civil war in 1949, retired from the army in 1954, and was doing "Party work" when he was assigned to help establish the Yingkou County Earthquake Office in September 1974. He lived in the town of Dashiqiao, where the county government was located.

China had launched a seismic prediction program in 1966 with 17 major centers nationwide, 250 seismic stations, and 5,000 observation points. Guided by scientists, more than 100,000 citizen observers were collecting data. Cao worked enthusiastically at the heart of this campaign and took his duties very seriously.

In keeping with Chairman Mao's ideology and his distrust of elites—including, to some extent, scientists—these citizen observers became the backbone of the program. While trained experts monitored seismographs and carefully surveyed markers across fault lines to track the build-up of stress in the rocks, geomagnetic anomalies, and the release of radon gas, "the masses," led by men like Cao, kept track of the fluctuations in well water and sometimes the strange behavior of animals, which according to tradition were symptoms of a coming disaster. A few local citizens were trained to measure basic changes in electrical currents in the ground.

After three centuries of relative calm, northern China had been rattled by a series of three large temblors between 1966 and 1969. While geologists in the West were just coming to terms with plate tectonics, earth scientists in China had not yet accepted the theory (and would not until after the Cultural Revolution), so it was only later that they realized the northeastern part of the country was being squeezed from the east and west by colliding plates in the Himalayas and the ongoing subduction of the ocean floor along the Japan–Kuril trench.

On June 29, 1974, the State Seismological Bureau issued a warning that earthquakes might occur within the next two years in an area that included the Yingkou area, home of Cao Xianqing. This was based on a torrent of new data that showed ground movement along the Jinzhou fault. There were also reports (never confirmed) of a rise in sea level in the area.

By mid-December things began to accelerate. Reports of radon gas in the wells emerged, along with changing water levels. Then snakes started crawling out of their hibernation dens, freezing to death on the snowy ground. And the ground had begun to shake. After a swarm of small tremors there came a series of high alerts predicting magnitude 4–5 shocks over the weeks ahead.

Less than a month later, at the next national quake prediction conference in Beijing, Gu Haoding, a seismologist from Liaoning's provincial Earthquake Office, drew attention to the short-line leveling data across the Jinzhou fracture and said that "rocks near the fault had reached the unstable stage of plastic deformation" and were on the verge of rupture. "Therefore, a relatively large earthquake will not be very far and should be in the first half of this year or even January and February."

Based on this leveling data from the Jinxian Observatory—in one case a 0.1-inch (2.7 mm) elevation change over a period of only ten days—plus the radon gas reports, a series of magnetic anomalies, and a rash of smaller tremors, Gu used an empirical formula to predict that the coming quake would be a magnitude 6 and would occur somewhere around the southern tip of the Liaodong Peninsula. Because radon gas emissions were reported as far inland as Panjin and other anomalies were indicated in Dandong and Gaixian, Gu decided to expand the potential danger zone to include the entire Liaodong Peninsula and its offshore regions. When challenged about his very short time estimate— within six months, although possibly as early as January or February—

Gu dug in his heels and said it could happen "even before the end of this conference."

When he turned out to be right—the quake struck two weeks later—Gu's report would retroactively be proclaimed the official "short-term" prediction of the Haicheng earthquake.

At the time of the mid-January conference in Beijing, though, higher-ups were not so sure. They decided to soften Gu's warning with a significant fudge factor. The final statement changed his prediction from a magnitude 6 within six months—possibly even two months—to a more vague magnitude 5 to 6 some time "in this year." After the late December false alarms, the sense of urgency had started to fade. Rumors had spread that the filling of a nearby reservoir was the likely cause of the swarm of smaller tremors in December, so life in northeastern China started to return to a semblance of normalcy. The seismic threat momentarily fell off the radar for some public officials.

Oddly enough, the number of reports of bizarre animal behaviors *increased* during mid- to late January. In the month before the quake more than a hundred snake sightings were recorded. What did modern-day scientists think of the snake stories? Kelin Wang and his coauthors commented that "although they must represent a tiny fraction of the total snake population in southern Liaoning, such suicidal behavior is extremely difficult to explain. What the snakes and other animals sensed is not known. It could be as simple as vibrations caused by earthquake tremors that were not detected by the then very sparse seismic network."

Along with the snakes, frogs were now coming out of their hibernation dens and freezing to death. Horses, chickens, and cows were making a racket that was duly noted by the cadre of amateur observers. Rats and mice were running around, seemingly disoriented, oblivious to cats and unafraid of people.

According to ancient Chinese lore, all of these animal anomalies were considered portents of a coming shock and people took them seriously, reporting the incidents to various observatories and government offices where men like Cao Xianqing jotted them down in logbooks. In the month before the quake he cranked out sixteen issues of his "Briefing Notes" to the county government. Cao had taken to heart a book about the Ningxia event of January 3, 1739. The book spelled out a series of circumstances and events that Cao could see unfolding before him every day of January 1975.

One significant passage apparently convinced him the time was near. "Earthquakes occur mostly in winter or spring. If well water suddenly turns muddy, there is lasting cannonlike sound from the ground, gangs of dogs bark furiously, one should be mindful of earthquakes. Excessive autumn rain will surely be followed by a winter earthquake." Cao knew it had rained a lot in the fall of 1974. In 1975, according to the Chinese calendar, February 4 would be the last day of winter and thus Cao decided there was no time to waste.

In late January, with disaster fever on the wane in some circles, Cao ramped up his own activities. While others were breathing a sigh of relief about the false alarms, he used his position as head of the Yingkou County Earthquake Office to accelerate preparations for the coming disaster. According to an entry in his logbook on January 28, Cao had already organized a twenty-one-person rescue team, a sixteen-person transportation team, twenty-five thousand kilograms of baked foods, a thousand winter jackets, ten thousand pairs of winter shoes, a thousand winter hats, and a thousand cotton quilts in anticipation of a large wintertime earthquake.

From his office in Dashiqiao, Cao maintained frequent telephone contact with the nearby Shipengyu Earthquake Observatory, where a seismograph and a tiltmeter were being monitored day and night. On the first two days of February, several small tremors—too small to be felt by people living in the area—were detected by the team at Shipengyu and while these minor tremors did not cause any alarm at first, the amplitude of the shockwaves seemed to increase with each new event. They were briefly noted in the station's logbook and Cao kept track of the rumbling as well.

Then a sudden surge of tectonic vibration began on the evening of February 3 and from that moment on things began to happen in rapid succession. A series of small rumbles bloomed into a rash of tremors. This burst of seismicity truly did alarm the workers at Shipengyu, who began a flurry of phone calls to notify the Party Committee, the army, city and county officials, and the Liaoning provincial Earthquake Office to "enhance preparation" for a larger quake.

At the same time they fielded a steady stream of incoming reports from communes and amateur observers with details of tumbled chimneys, fallen gables, and broken windows. Cao himself called Shipengyu to report that water in a dozen wells had dropped twelve to twenty inches (30–50 cm) during the evening, that several other wells that had contained water during the day were now completely dry, and that horses and chickens were making a lot of noise and trying to escape.

By eight o'clock the next morning—February 4, 1975—there had been two hundred tremors, culminating in a magnitude 5.1 event at 7:51 a.m. At 8:15 a.m., seven members of the Party Committee of Yingkou County held an extended emergency meeting at Cao's urging. When he summarized the overnight flurry of tremors and damage reports and announced that a large earthquake "may occur today during the day or in the evening," he must have made a convincing case. Immediately after the meeting the committee—without waiting for approval or instructions from higher authority—issued its own sternly worded statement canceling all business and production work, all public meetings, all entertainment, and all sporting activities immediately.

Miles away at the Liaoning provincial Earthquake Office a simultaneous 8:00 a.m. meeting was taking place in which head scientist Zhu Fengming explained the worrisome overnight data to higher-up Party officials. In a bulletin issued just after midnight, Zhu had cautioned them that "the magnitudes are still increasing" and "a relatively large earthquake is very likely to follow." Nothing quite this urgent had ever been written in previous reports at the provincial level. But the tone was still several notches below the intensity of alarm that Cao Xianqing was raising with local county officials in Dashiqiao.

When Kelin Wang interviewed Zhu and his colleagues in 2004, the senior scientist and his fellow workers in the provincial office confided that none of them had attempted—or felt they were able—to predict a rupture on a given day. By "very likely to follow," Zhu told Wang he meant "a timeframe of one to two weeks." Basically, Zhu was the cautious scientist and Cao was more the gambler. Cao turned out to be right, but he never could explain how his prediction was made. Despite the contrast between Zhu's conservative warning and Cao's bold campaign to evacuate, Chinese officials would later credit Zhu and his provincial colleagues with having made the "imminent prediction" of the Haicheng earthquake.

The Revolutionary Committee of Liaoning province organized a ninety-minute meeting for two o'clock that afternoon in the Haicheng guesthouse, to be attended by a dozen government officials from Haicheng and Yingkou Counties—Cao Xianqing among them—along with an army officer from the People's Liberation Army troop that was deployed in the area. During the meeting Li Fuxiang from the provincial Earthquake Office estimated a magnitude 6 or greater "may occur within the next few days." Again, not quite the strong message that Cao was delivering in his hometown. While he attended the somewhat inconclusive two o'clock meeting, the evacuation of buildings in Dashiqiao and Yingkou County was already well underway.

Throughout the afternoon Cao continued to announce that a large shockwave would occur *that day*. Word was spread by telephone calls to communes and production brigades and by loudspeaker broadcasts in the streets. When the swarm of small tremors recorded at Shipengyu hit 501—and then went quiet—late in the afternoon, Cao's intuition told him this was the calm before the storm, the final energy build-up before the big rupture. He was heard to say that the later the quake occurred the larger it would be, a magnitude "seven at seven o'clock and eight at eight o'clock." Hearing this, some of the more senior scientists smirked at Cao's homespun certainty.

Meanwhile, Shipengyu Earthquake Observatory workers had been spreading the word as well. They convinced the movie operator of the nearby Shipengyu Production Brigade that a jolt was coming *that night*, so word went out that there would be movies that evening. Movies in the village were always projected outdoors. The hope was that people would be attracted away from their houses—another spontaneous, local decision that definitely saved lives.

Cao and his colleagues faced another, more daunting challenge that evening. With the Chinese New Year approaching on February 11, the city of Anshan had dispatched a leading cadre and a greeting delegation to Cao's hometown of Dashiqiao to express good wishes and to give a stage performance to entertain a headquarters delegation from the 39th Army stationed there.

Because of the prediction, army officials had to decide whether or not to go ahead with the evening's festivities, scheduled for 7:00 p.m. Simply canceling was not an option since the Anshan delegation was already in town. The ranking army commander was said to be furious about the situation. He decided to cancel the stage performance, but he did insist that the greeting ceremony proceed as planned—a sign of respect for the high-level delegates who had traveled all the way to Dashiqiao. Just to be on the safe side, all seven doors of the assembly hall were kept wide open on a cold winter's night.

Come seven o'clock, Anshan's leading cadre delivered a mercifully brief speech, the army brass expressed their gratitude, and the whole affair was wrapped up by 7:20 p.m. In keeping with protocol, high-ranking officers and the cadre who sat on stage left the building first. Nearly a thousand in the audience patiently followed in an orderly fashion, the last few walking out the door at exactly 7:36 p.m.—just as the magnitude 7.3 Haicheng earthquake began. The assembly hall collapsed in a heap of rubble.

The rupture occurred near the boundary line between Haicheng and Yingkou Counties, a left-lateral slip on a northwest-trending "blind" fault (one that did not show at the surface) that was not known to exist before the earthquake. It did not occur on the southern peninsula along the segment of the Jinzhou fracture zone they'd been watching so closely. The epicenter turned out to be 125 miles (200 km) northeast of there— between Dashiqiao and Haicheng—only twelve miles (20 km) away from the Shipengyu Earthquake Observatory. The shockwaves caused extensive ground failure and liquefaction in both counties, with widespread destruction to buildings and infrastructure in dozens of villages and towns in a rural–urban region with a population of roughly a million.

Early casualty and damage reports were both convoluted and classified. For some reason it was customary in China to record living space in rural communities by the number of individual rooms but in urban centers by square area (including schools, offices, and factories). So rather than count the total number of buildings wrecked, a secret document on the Haicheng incident reported 17,497,342 square yards (14,630,000 m²) of urban living space and 1,840,000 rural rooms had been damaged or destroyed. More than 6.5 million feet (2 million meters) of various transportation lines and pipelines were damaged, more than seven hundred hydraulic facilities and two thousand bridges were damaged, and seventy square miles (180 km²) of farmland were wrecked by liquefaction and sand fountains. The total economic loss was estimated to be at least a billion yuan.

The reported death toll, a number both political and vague, was also kept secret until after the end of the Cultural Revolution. In the immediate aftermath phrases like "a few fatalities" in a population of a million people appeared in press releases from the official Chinese news agency. In 1988, when a Chinese researcher with better access to original documents reported a death toll of 1,328, skeptics in the West immediately pounced on the so-called discrepancy as evidence that the whole story of the Haicheng prediction was a gross exaggeration.

One skeptic in particular, Robert J. Geller, an American seismologist on the faculty at the University of Tokyo and a vehement doubter of the value and practicality of seismic forecasting worldwide, commented that "the large disparity between the reports of 1975 and 1988 casts doubt on claims for the Haicheng prediction." Kelin Wang and his colleagues in 2004 pointed out that there was no real discrepancy. It was simply a case of China not releasing the early details to foreigners in the 1970s.

In March, a month after the rupture, Party officials launched a publicity campaign with a news release that said, "The earthquake-work team of our country predicted this earthquake; under the unified leadership of the Liaoning Provincial Committee of the Chinese Communist Party, the Party [members], government, army, and masses in the epicentral area took timely and effective preventative measures, so that losses caused by the earthquake in this densely populated area were greatly reduced. This is a vivid demonstration of the superiority of our country's socialist system. This is a great victory of Chairman Mao's proletarian revolutionary line!"

In other words the exact number of deaths didn't really matter. What counted, in the minds of Chairman Mao's publicity machine, was that China's ability to mobilize the masses to tackle a complex scientific challenge had paid off. A natural disaster that could have killed many more did not.

When Kelin Wang and company were allowed to see previously secret documents in 2004, a new and presumably more accurate set of numbers could be compiled for the first time. Combining "direct" and "indirect" causes—many died from fires and as many as 372 died from hypothermia, freezing to death outside just like the snakes—the total death toll for the Haicheng earthquake became 2,041 and the total number of people injured was 24,538.

To make the evacuation story look even more impressive, some news releases described the total region affected as having more than eight million people. Even if the population in the epicentral area was really closer to a million—the original figure cited by most authorities—at 2,041, the number of deaths is still quite small given the collapse of so many houses. A report by the U.S. delegation of scientists who visited the disaster zone in 1977 estimated that "casualties in excess of 100,000 would have ordinarily been anticipated."

It's safe to say the Chinese government had a success story worthy of bragging about even without fudging the numbers. The political turmoil of the late 1970s, however, affected the publicity machine that took over in the wake of Haicheng. The context, as Kelin Wang explained it, was that "with Chairman Mao's health deteriorating, friction between the Gang of Four and other Party leaders, including Mao's would-be successor, Hua Guofeng, intensified." Mao's wife, Jiang Qing, and her three allies in the Gang of Four were running the publicity campaign, but Hua Guofeng had his own agenda and an element of the truth may have been the first casualty.

As vice-premier, Hua was among the first to heap praise on the workers based at the Shipengyu Earthquake Observatory for their stellar efforts to warn the community about the rash of foreshocks—now known to have been true precursors of the big jolt. Followers of the Gang of Four, on the other hand, issued news releases that stressed the leadership role of provincial *Party* officials with little or no mention of Shipengyu, much less of Cao Xianqing. To make the most of this singular propaganda opportunity, it was evidently better to avoid the devilish details.

The prediction was real yet the results of the evacuation were uneven. Yingkou County's warning was spread far and wide from early that morning, thanks to the single-mindedness of Cao and his team. In Haicheng County the warnings began later in the day, many decisions to evacuate were made spur of the moment by local committees or individuals at the work brigade or commune level, and some parts of the province—including the town of Haicheng itself—were not evacuated at all. The death toll in Haicheng County was substantially higher than in Yingkou. Thus it became a question of who got the greater credit, or who deserved the blame.

Three months after the disaster, when Vice-Premier Hua delivered a speech at the next national quake prediction conference, he told the famous story of the evacuation of delegates from the meeting hall that night in Dashiqiao just as the ground started to shake. "An officer who was directing people to exit was injured," said Hua, "but the rest of the one thousand people were all safe." And who made the prediction the army commanders took so seriously, the prediction that saved so many lives? Hua Guofeng said it was the Shipengyu Earthquake Observatory. The Gang of Four, had they attended the conference, would no doubt have touted Party bureaucrats.

Cao Xianqing insisted it was his Earthquake Office that provided the vital first warning. No doubt the army brass heard about it from several sources, but Cao's prediction turned out to be the most specific and timely. Perhaps because he was not a trained scientist and could not provide a detailed technical explanation for his decisions—how much was science and how much was instinct?—or perhaps because he was a lowly county official rather than a provincial bigwig, he did not get anointed as a hero of the Haicheng saga. In the feuding between Hua and the Gang of Four, Cao received no official recognition, no meritorious service awards. When China bragged about the world's first successful seismic prediction and evacuation, the story of Cao Xianqing and his Yingkou County Earthquake Office was not told to the outside world.

The lingering question was and still is whether the prediction had any scientific merit. When Kelin Wang and his colleagues published their findings in 2006, they found that "the most important precursor was a foreshock sequence, but other anomalies such as geodetic deformation, changes in groundwater level, color, and chemistry, and peculiar animal behavior also played a role." In essence, they wrote, "None of these predictions can be scientifically explained."

The point seemed to be that even though some things like why snakes crawled out of their dens or why some—not all—mice and rats were dazed and disoriented may not have clear scientific explanations, there was no reason to doubt that they really did happen. Wang and company confirmed that "it was the foreshocks alone that triggered the final decisions" to warn and evacuate. The good news was Haicheng proved that "at least some earthquakes do have precursors that may lead to some prediction." To me the bottom line was this conclusion: "Although the prediction of the Haicheng earthquake was a blend of confusion, empirical analysis, intuitive judgment, and good luck, it was an attempt to predict a major earthquake that for the first time did not end up with practical failure."

Seventeen months and three weeks after Haicheng, everything China thought it knew about seismic prediction came down like a house of cards. At 3:42 a.m. on July 28, 1976, lightning flashed across the sky and the earth rumbled ominously. Seconds later another major earthquake struck northern China—this time with no prediction and no evacuation. The rupture happened directly underneath the industrial city of Tangshan, roughly 90 miles (140 km) east of Beijing.

The magnitude 7.5 rupture shifted the ground about five feet (1.5 m) horizontally and three feet (1 m) vertically, destroying nearly 100 percent of the living quarters and 80 percent of the industrial buildings in the city. People were jolted from their sleep in total darkness, screaming and choking on thick dust from unreinforced brick buildings that collapsed in piles of rubble. Official reports estimated the death toll at 240,000 with 164,000 more seriously injured. That's roughly the same number who died from the Sumatra quake *and* the Indian Ocean tsunami in 2004. Critics claimed the Tangshan estimate was conservative, however, and that the real number of people killed could have been 600,000 according to estimates made by foreign observers. Whatever the count, Tangshan was far and away the most deadly single quake in the twentieth century and one of the great tragedies of all time.

Tangshan was a major center for coal mining, iron and steel production, and the manufacture of cement. Nearly all of it was wrecked. Bridges and highways collapsed, pipelines broke, dams cracked, more than ten thousand large industrial chimneys fell, and twenty-eight trains passing through the city overturned or were derailed. The key question, though, was why nobody saw it coming. Did the lessons of Haicheng not apply here? Apparently not.

In the final two months before Tangshan, not a single foreshock was detected by a regional seismic network capable of measuring tremors as small as magnitude 1.7. Some of the other little twitches and anomalies that had preceded the Haicheng event also preceded Tangshan but apparently the signals were not strong enough to trigger a prediction or evacuation. How different could the geological structures be only three hundred miles (480 km) away from Haicheng? If Haicheng had foreshocks, why not Tangshan? Good and important questions that still need to be answered, said Kelin Wang.

Preliminary answers suggested by Wang and company could be that fault failures are unique, so different from each other that whatever anomaly or precursor helps to predict one event probably won't work for a different kind of fault in a different physical setting. And once a rupture has happened, it modifies the geological structures and rock properties enough that the precursors then change as well. The symptom that tipped us off to the *last* quake may not precede the next, even if it happens on the same fault. And then sometimes you get the symptoms but the big temblor never comes.

If that's the real lesson of Haicheng and Tangshan, why bother with prediction? When I finally got a chance to interview Kelin Wang, he still seemed hopeful about the prospects—a persistent, low-key optimist—although he was wary of putting prediction into practice too soon. "If the Haicheng story was true," he began, "then earthquake prediction is not impossible." It did show us "how *early* in the stage we are still in terms of earthquake prediction. And the difference—the major difference—between these two earthquakes is the Haicheng earthquake had a foreshock sequence and the Tangshan earthquake had no foreshocks at all." Strange as it may sound this did little to dampen enthusiasm for the dark art of prediction.

The Tangshan disaster gave prediction optimists a sharp reality check. So did the Parkfield experiment in California. On September 28, 2004, a magnitude 6 temblor finally rattled the farming town of Parkfield—at least twelve years after scientists predicted it would happen. To say that William Bakun and Allan Lindh of the USGS, who along with Tom McEvilly of the University of California at Berkeley had offered the forecast in 1985, were disappointed would probably be an understatement. Five moderate (magnitude 6) events with similar "characteristics" had occurred on the Parkfield segment of the San Andreas since 1857. By their calculations the seventh in what looked like a repeating series of nearly identical ruptures should have happened some time in 1988 but surely by the end of 1992 with a 95 percent probability.

Clearly the "time-predictable" part of their hypothesis was wrong. The idea that fault failures tend to repeat themselves like clockwork had been kicking around since 1910, when geologist Harry F. Reid of Johns Hopkins University suggested it ought to be possible to figure out when and where quakes would happen by keeping close tabs on the build-up of stress. Looking at how unevenly land had shifted along the San Andreas during the great San Francisco earthquake of 1906, Reid developed the elastic rebound hypothesis—a cornerstone of modern geology long before the advent of plate tectonics—which Bakun, Lindh, and McEvilly set out to test in Parkfield eight decades later.

Reid's idea was that stress built up unevenly along the fault and it took a massive rupture—at the point where the strain was great enough to cause the rocks to fail—in order to relieve or *recover* that strain. The longer the strain built up, the bigger the shock would be. If you knew how often the fault had ruptured in the past, you could in theory estimate how long before the next one was due.

But what if the last rip did not release all of the accumulated strain? Wouldn't that alter the timeline for the next one? When Bakun and Lindh published their forecast in the August 16, 1985, edition of *Science*, they noted that the 1983 magnitude 6.5 at Coalinga, eighteen miles (30 km) off the San Andreas to the northeast of Parkfield, might have done exactly that. It had just possibly relieved enough of the "tension in the spring" of Parkfield's clock to delay the next rumble in the series. A delay of more than a dozen years, however, was way more than merely late.

Critics within the science community didn't wait until the 2004 jolt to pounce on Parkfield. Even though the expected magnitude 6 event did happen *eventually*, in more or less the same location as last time and where Bakun and his colleagues said it would be, the Parkfield prediction experiment was branded a failure shortly after the original time window closed in January 1993. A long-standing and rancorous philosophical debate intensified as some seismologists turned away from divining the future and deleted "the P-word" from their vocabularies.

By the mid-1990s Robert J. Geller at the University of Tokyo had become the most persistent and outspoken critic of everything predictive, especially Japan's massive and well-funded multiyear effort to anticipate the next big temblor near Tokyo. Geller had been scathing in his view of American efforts as well, his central thesis being that prediction studies have been going on for more than a century—and yet we seemed no nearer a solution to the problem than we were in the beginning.

Geller was fond of quoting Charles Richter, developer of the earliest and best-known earthquake magnitude scale and one of the most respected seismologists in the world, who in 1977 commented that he had "a horror of predictions and of predictors. Journalists and the general public rush to any suggestion of earthquake prediction like hogs toward a full trough." Vitriol and aspersions aside, Geller's central argument against prediction was and is based on the idea that "individual earthquakes are inherently unpredictable because of the chaotic, highly nonlinear nature of the source process." Basically there are so many things going on deep underground that we can never know when a rock surface is going to fail. He dismissed the idea that "the Earth telegraphs its punches," using auto accidents as an analogy.

The rate of car crashes may be estimated, but the time and location of an individual accident "cannot be predicted." As for precursors, even though speeding frequently precedes accidents, only a small fraction of speeding violations are followed by serious accidents. Therefore speeding is not a reliable precursor. Similarly, he argued, there are no reliable precursors to seismic shocks. Even after a car crash has begun to happen, its final extent and severity depend on other equally unpredictable, quickly changing interactions between drivers, cars, and other objects. Put simply, car wrecks and quakes are too chaotic to foretell, according to Geller.

In October 1996 he joined forces with David Jackson and Yan Kagan at UCLA and Francesco Mulargia at the University of Bologna to write a critique for *Science* that appeared under the provocative headline "Earthquakes Cannot Be Predicted." In the article they cast doubt on the Haicheng prediction story, suggesting that political pressures had led to exaggerated claims. They wrote that there are "strong reasons to doubt" that any observable, identifiable, or reliable precursors exist. They pointed out that long-term predictions both for the Tokai region in Japan and for Parkfield had failed while other damaging jolts (Loma Prieta, Landers, and Northridge in California, plus Okushiri Island and Kobe in Japan) had *not* been predicted. They cautioned that false hopes about the effectiveness of prediction efforts had already created negative side effects.

After the frightening and damaging Northridge temblor in southern California, for example, stories began to spread that an even larger quake was about to happen but that scientists were keeping quiet to avoid causing panic. The gossip became so widespread that Caltech seismologists felt compelled to issue a denial: "Aftershocks will continue. However, the rumor of the prediction of a major earthquake is false. Caltech cannot release predictions since it is impossible to predict earthquakes."

Not surprisingly the article spawned a series of energetic replies from those who felt the baby of prediction science should not be thrown out with the bathwater of uncertainty. Max Wyss at the University of Alaska took issue with almost every point made by Geller and company. He countered that in 10 to 30 percent of large quakes foreshocks do occur and are precursors, that strain is released in earthquakes only after it has been accumulated for centuries, and that measuring the build-up of stresses within the crust is therefore not a waste of time and money.

Wyss concluded that most experts living at the time of Columbus would have said it was impossible to reach India by sailing west from Europe and that "funds should not be wasted on such a folly." And while Geller et al seemed to be making a similar mistake, Wyss doubted that "human curiosity and ingenuity can be prevented in the long run." The secrets of quake prediction would be unlocked sooner or later. Richard Aceves and Stephen Park at University of California Riverside suggested it was premature to give up on prediction. "The length of an experiment," they wrote, "should not be an argument against the potential value of the eventual results."

In a later article Geller repeated his contention that "people would be far better off living and working in buildings that were designed to withstand earthquakes when they did occur." He insisted that the "incorrect impression" quakes can be foretold leads to "wasting funds on pointless prediction research," diverting resources from more "practical precautions that could save lives and reduce property damage when a quake comes."

∞

In the spring of 1997 someone with inside knowledge leaked a government document that slammed Japan's vaunted $147 million a year prediction research program. The confidential review, published in the *Yomiuri Shimbun,* quoted Masayuki Kikuchi, a seismologist at the University of Tokyo's Earthquake Research Institute, as saying that "trying to predict earthquakes is unreasonable." After thirty-two years of trying, all those scientists and all that high-tech equipment had failed to meet the stated goal of warning the public of impending earthquakes.

The report said the government should admit that seismic forecasting was not currently possible and shift the program's focus. It was the sharpest criticism ever, and it did eventually lead to a change in direction. With so much invested and so much more at stake, though, there was no way the whole campaign would be ditched. People in Japan are intimately aware of earthquakes and the public desire for some kind of warning—whether *unreasonable* or not—is a political reality that cannot be ignored.

Faults in the Tokai region off the coast of Japan—where *three* tectonic plates come together—have rattled the earth repeatedly and people worry about the next one. The subduction zone there tore apart in 1854, the great Tokyo earthquake of 1923 killed more than 140,000 people, two more big fault breaks occurred in the 1940s, and another magnitude 8 is expected any day now.

Japan's first five-year prediction research plan was launched in 1965. In 1978, with still no sign of an impending quake, the program was ramped up with passage of the Large-Scale Earthquake Countermeasures Act, which concentrated most of the nation's seismic brain power and technical resources on the so-called Tokai Gap. Whenever some anomaly is observed by the monitoring network, a special evaluation committee of technical experts—known locally as "the six wise men"—must be paged and rushed by police cars to a command center in Tokyo, where they will gather around a conference table and focus on the data stream. Then very quickly they must decide whether or not to call the prime minister.

If the anomaly is identified as a reliable precursor, only the prime minister has the authority to issue a warning to Tokyo's thirteen million residents. If and when that day comes, a large-scale emergency operation will be initiated almost immediately. Bullet trains and factory production lines will be stopped, gas pipelines will shut down, highway traffic will be diverted, schools will be evacuated, and businesses will close. According to one study a shutdown like that would cost the Japanese economy as much as $7 billion per day, so the six wise men can't afford to get it wrong. False alarms would be exceedingly unwelcome.

Even though the people of Japan still tell their leaders they want some kind of warning system, they were not at all impressed with what happened in Kobe in 1995. With all those smart people and so much equipment focused on the Tokai Gap, apparently nobody saw the Kobe quake coming. It was an ugly surprise from a fault that was not considered a threat. It killed more than six thousand people.

In spite of the embarrassing setback, Kiyoo Mogi, a professor at Nihon University and then chair of the wise men's committee, defended the prediction program, calling it Japan's moral obligation not only to its own citizens but to people in poorer, quake-prone countries around the world as well. "Can we give up efforts at prediction and just passively wait for a big one?" he asked. "I don't think so." What Mogi did was try to change the rules.

He argued that a definite "yes or no" prediction—as the six wise men are required by law to make—was beyond Japan's technical capability with the knowledge and equipment available. Instead, he suggested the warnings be graded with some level of probability, expressed like weather forecasts. The government could say, for example, that there's a 40 percent chance of an earthquake this week. People would be made aware that it *might* happen and that they ought to prepare themselves.

Mogi's idea was rejected, so he resigned from the committee in 1996. The program carried on but it gradually changed direction. In the aftermath of Kobe prediction research spending actually increased again with installation of a dense web of GPS stations to monitor crustal movement and strain build-up. But by September 2003, with all the new equipment up and running, an array of 1,224 GPS stations and about 1,000 seismometers failed to spot any symptoms of the magnitude 8.3 Tokachi-Oki earthquake. It came as another rude shock.

The prediction team had also started work on what they called a "real-time seismic warning system." Japanese scientists were hoping to use super-fast technology to reduce the extent and severity of damage once a fault had begun to slip. They loaded a supercomputer with 100,000 preprogrammed scenarios based on the magnitude and exact location of the coming temblor. As soon as the ground began to shake instruments would feed data to the computer and the computer would spit out the most likely scenario—one minute later.

But on June 13, 2008, a magnitude 6.9 shockwave hit northern Japan, killing at least thirteen people, destroying homes and factories throughout the region. The real-time system did signal that a powerful jolt was happening—roughly three and a half seconds after it started—but the source of the quake was too close to be of any use to places like Oshu, which was only eighteen miles (30 km) from the epicenter. People there received 0.3 seconds of warning. The unfortunate reality is that those closest to the strongest level of shaking will always be the ones who receive the shortest notice. Even if the system works exactly as it should, a real-time warning system will benefit primarily those farther away. On the other hand, it could stop or slow the spread of fires and speed the arrival of emergency crews. So in Japan—at least for the foreseeable future—the supercomputer and the six wise men still have a job to do.

∞

Not only was the Parkfield earthquake a dozen years late but the densely woven grid of seismographs, strainmeters, lasers, and other equipment that made the area one of the most closely watched rupture patches in the world had apparently failed to spot any obvious symptoms or definite precursors. In 1934 and 1966 the Parkfield main shocks had been preceded by apparently identical magnitude 5 foreshocks, each about seventeen minutes prior to the magnitude 6 main event. But not this time.

In 1966 as well, the fault had seemed to creep a bit more than normal in the weeks before the failure. There were reports of new cracks in the ground and a water pipe crossing the zone broke the night before the rupture. Nothing like that happened before the 2004 event. No obvious foreshocks or slip before the main event. Seven "creepmeters" were deployed along the rupture zone with nothing to show for the effort. But all was not lost according to Allan Lindh, who in early 2005 wrote an opinion piece for *Seismological Research Letters* defending the work at Parkfield. His paper sounded a new rallying cry for prediction science.

Looking closely at *where* the break occurred, how strong it was, and the aftershock pattern that followed, he argued that a key part of their original prediction *had* come true. What happened in 2004 was physically "a near-perfect repeat" of the 1966 event, according to Lindh. The same earthquake happened again—rupturing the same fifteen-mile-long (25 km) segment of the San Andreas between the same two little bends or "discontinuities" in the rock and with the same overall magnitude—what some had called Parkfield's signature or "characteristic" earthquake. One might have expected the magnitude to be greater than 6 because the jolt was twelve to fifteen years later than expected and therefore had more time to accumulate strain in the rocks. But the magnitude was 6, just like its predecessors. Hence, Lindh argued, it was a repeat of the same event.

One curious twist was that the 1966 event had ripped the fault from north to south while this time it unzipped from south to north. And according to Lindh, there may have been a "small premonitory signal" at three or four Parkfield strainmeters. Holes had been drilled hundreds of feet down into the fracture zone in the late 1990s and extremely sensitive instruments capable of detecting very small increments of stress had indeed recorded signals "of the order of 10 nanostrain," if all the devices were working properly.

While this sounded like an infinitesimally small thing to measure, Lindh pointed out that if the coming quake had been a magnitude 7 instead of a 6, then the amount of strain—and the creep along the fault—would probably have scaled up by a factor of ten, which "would be easily observable with current downhole instrumentation." His point was that new state-of-the-art strainmeters can detect things even those magical GPS rigs cannot see from above.

When a fault creeps way down deep, not all the horizontal motion is transferred to the surface because rocks bend and deform under stress. Therefore, if the rocks started to move hundreds of feet below ground, and if this turned out to be a reliable symptom of a coming rupture, the GPS stations up at the surface might not detect the signal even at the supposed magnitude 7 level. But the much more sensitive strainmeters *could*—or might.

Lindh took the opportunity to bite back at critics of prediction and those in government committees, labs, and universities who had unofficially given up trying to solve the problem. He ridiculed the fashion in California of making "probability forecasts" so vague that the results were "no longer accurate enough" to be of use to society. For example, the official Working Group on California Earthquake Probabilities has estimated the odds of at least one magnitude 6.7 or larger event in the San Francisco region sometime in the next thirty years to be 62 percent. The odds are 67 percent for the same size jolt to hit the Los Angeles area. But sometime in the next thirty years? How should people respond to a prediction like that? Building codes and insurance rates might be adjusted, but how do the rest of us make sense of it?

The numbers could mean anything or nothing to the majority of citizens, who don't really comprehend statistical probabilities. Instead Lindh argued that the data collected at Parkfield and elsewhere in recent years had vastly improved our understanding of fault behavior, of the accumulation of strain, and of seismic patterns over longer periods so that it should be possible now to target the three most likely zones to rupture and to design more focused "prediction experiments" that could save lives.

"While I understand that some in the field of seismology are afraid of the 'P word,' the public is not," wrote Lindh. "They think it's what seismologists are working on. It is my opinion that the public would respond very positively to our highlighting some of the most serious threats to their lives and welfare, particularly if it were accompanied by a serious commitment to do everything we could to further our understanding of those segments, and maybe in the process even reduce the risk they represent."

Prediction fell from grace with a disappointing thud just as Kazushige Obara in Japan and Herb Dragert and Garry Rogers at the Geological Survey of Canada were learning about ETS (episodic tremor and slip), the bizarrely regular twitching deep down on the lower reaches of the Cascadia Subduction Zone that promised a new way to track the behavior of a major fault. At roughly the same time in California, all that high-tech equipment buried in the ground or stretched across the San Andreas at Parkfield—the creepmeters, the tiltmeters, the seismographs, and lasers—was being reconfigured for a new and bigger experiment. The desire to know more about what happens along the rocky surfaces of a fracture zone just *before* an earthquake was actually gaining momentum despite the alleged failure of the Parkfield prediction.

Parkfield was reborn as SAFOD—the San Andreas Fault Observatory at Depth, a deep-earth research project funded by the National Science Foundation in partnership with the USGS. In June 2004 an oil rig crew started drilling a hole into the hilly brown rangeland not far from the initiating point of the 1966 Parkfield temblor. With a rotary bit nearly ten inches (25 cm) in diameter, they sank a shaft almost two miles (3 km) into the earth on the western side of the fault and installed another package of instruments designed to monitor the initiation of small, repetitious earthquakes at close range.

At the end of September, when the long-awaited Parkfield event finally happened, there was no immediate payoff because, as even Bakun and Lindh admitted, the new equipment did not detect any obvious precursors. But the drilling continued, enthusiasm undimmed. The next phase of the SAFOD project would deploy the oil industry's newest directional-drilling technology to turn the bit almost sideways and then drill *through* the fault—from the Pacific plate eastward—until it penetrated the gap and reached relatively undisturbed rock in the North America plate on the eastern side of the San Andreas. The plan was to bring up core samples of rocks and fluids to find any secret ingredient that might cause ruptures to begin.

The project would also implant more instruments inside the active zone to make long-term measurements of small to moderate tremors and continuous measurements of rock deformation as it built up during the next cycle. Nothing on this scale had ever been tried. These were ambitious goals: dig down to the very heart of an active fault and watch it rupture from the inside.

SAFOD was only one of three components of an even grander science project called EarthScope, which set out to monitor plate tectonic movement along the entire U.S. west coast and create a 3D seismic image of the basement of North America. The Plate Boundary Observatory would do on a continental scale what SAFOD was doing close up on the San Andreas. The USArray—a spiderweb of new seismometers spun across the lower 48—would probe thousands of miles down to study the forces that create and shape the earth's crust from the bottom up.

To me it sounded like NASA gone underground. In fact, when I rang geophysicist and project director Greg van der Vink in Washington for some background, he volunteered his own analogy that Earth-Scope was geology's equivalent of a lunar landing, "the biggest thing we've ever done." But with a $200 million construction and installation budget for the first five years, EarthScope was really more like NASA on a crash diet, although it was still an impressive undertaking. And it would eventually become the Plate Boundary Observatory's job to focus a sharp new lens on Cascadia's fault.

With satellite technology that could measure plate movements down to half a centimeter, the system was intended to cover the western edge of North America from Mexico to Alaska with receivers spaced roughly 125 miles (200 km) apart. If the funding held out there would eventually be 875 permanent GPS stations working in concert with 175 deep borehole strainmeters 650 feet (200 m) underground to measure "at the proton level" what satellites cannot see from space. On standby would be another pool of a hundred *portable* GPS receivers for temporary deployment and rapid response to volcanic and tectonic emergencies.

Although EarthScope's budget looked flush by Canadian standards, it had taken van der Vink and others a long time to convince Washington politicians to spend money on something as optically unsexy as geology. Officials at the National Science Foundation insisted the money be spent in the United States. If Mexico and Canada wanted to join the project they would have to pay for their own equipment.

After the Rogers and Dragert findings were published in *Science*, however, news about episodic tremor and slip spread quickly. Greg van der Vink told me that ETS was "the poster child" for EarthScope science, exactly the kind of thing they were meant to study and "one of the most exciting new discoveries in a long time." As the implications sank in—here was a major fault sending some kind of mysterious signal every fourteen months like a giant metronome—the Cascadia Subduction Zone suddenly became a higher priority.

When Herb Dragert heard about the Plate Boundary Observatory project he sensed an opportunity. Having a few of those deep borehole strainmeters installed in Canada would be a great way to double-check his own findings. The Americans, realizing that if Canada could not afford to install strainmeters of its own there might be a huge gap in the data flow right at the most critical point along the locked zone where Cascadia's next rupture was most likely to happen, lobbied for an exception to the rule. Canada got the borehole strainmeters.

"Initially they were going to put six in the entire Pacific Northwest. From northern California to Vancouver Island—*six* strainmeters." Dragert sounded more than a little incredulous. "There's thirty-five now," he chuckled, "because they want to find out about ETS."

Like Dragert, Garry Rogers was especially keen to have another, independent set of instruments measure Cascadia's ground motions during the slip events just to make sure it was really happening. "When you start seeing phenomena with *several* kinds of instruments seeing the same thing, it becomes very convincing to a lot more people in the science world," Rogers explained. "In fact now three different kinds of measuring techniques—GPS, strainmeters, and seismometers—they're all telling us the same thing. And they're all telling us that stress build-up has a time element to it."

The old notion that stress build-up along the fault was a slow, steady, constant process caused by tectonic plates always moving against each other at roughly the same speed was apparently not accurate. Or the concept was more complex than early thinkers realized. The plates may be moving at a steady rate, but with an earth made up of all kinds of hard and soft rocks, mud, sand, and messy fluids, the build-up of stress between two plates is jerky.

"ETS events could be essentially like the clicks of a ratchet wrench," said Chris Goldfinger at Oregon State, continuing the thought. "As you crank it tighter and tighter, you're adding more and more load—as the Juan de Fuca plate tries to dive into the mantle. But the locking point between the two plates won't let it go, at this point, so it's *giving*—in small, squishy motions that may be cranking up the load for the big earthquake."

"You actually find that the probability of a megathrust earthquake is *larger* during one of these slips—or immediately *after* one of these slip events—than the rest of the time," ventured Dragert. According to calculations made by his GSC colleagues Stephane Mazzotti and John Adams, the risk jumps by a factor of about thirty. "Right now, roughly three hundred years into the cycle [Cascadia's last big quake was 311 years ago], the probability of a megathrust earthquake next week is roughly one in 200,000. So it's a very low probability. *During* the slip event, or immediately *after* a slip event, it's maybe twenty or thirty times that. It's still only one in four or five thousand, so it's still a low probability. But the difference is a factor of twenty or thirty."

To me the numbers or percentages seemed less significant than the idea that Cascadia's level of risk goes up for about ten days every fourteen months. And if it's true that a new load of stress gets shifted from lower down in the zone to the higher-up, locked part where the earthquake will eventually be generated, then it probably makes sense that one of these ETS events could eventually trigger the main event. Ruptures on Cascadia's fault (and the other subduction zones) may not be completely random after all. A new glimmer of hope for prediction optimists.

"That's what we've been looking for," said Dragert emphatically. "We won't call it a prediction yet, but I think once we know what the heck is going on here, we might be able to say, 'One of these slip events has started . . . The *probability* of a triggered event is higher than in the previous fourteen months.' What the emergency response people do with that—it's up to them." Dragert and Rogers have already suggested that emergency responders conduct their annual earthquake training exercises during ETS events just to raise awareness and have everybody thinking seismically during that zone of higher probability. Just in case.

"The ETS intervals are different in other places in the world," said Rogers, adding another complication. In northern California the cycle is much shorter than in British Columbia. In Oregon it's longer. "So that's something we don't understand," he said. "Why is it different from one subduction zone to another?"

With the setbacks at Parkfield still fresh in everyone's mind, it seemed to me unlikely that many of the skeptics would change their opinions and agree that quakes do come in time-dependent cycles. The discovery of ETS, though, did inject new energy into the prediction quest. Just when chaos theory seemed to have won the day, there was a fresh reason to think that at least some seismic shocks might not be a statistical crapshoot.

Garry Rogers was cautiously optimistic when I asked whether earthquake prediction had a future. He hesitated a moment, choosing his words carefully. "Under certain circumstances—yes," he said. "I think maybe Cascadia, maybe specific faults like the San Andreas that are very seriously studied, or specific faults in China—yes. I think I'm optimistic that that will happen. I'm not optimistic that we're going to be able to predict earthquakes everywhere. And I'm not optimistic that any of the predictions are going to happen soon." He smiled as I winced at all of the qualifications to his optimism. "So that's a qualified yes, if you like. We just don't know enough. It's a really tough problem."

Chris Goldfinger made a point of putting the Parkfield setback into a larger context. "Forecasting and prediction were words that were in great disfavor in the past couple of decades, partly based on the great Parkfield experiment," he said. "It wasn't really a failure—I don't think—at all. People expected too much from a one-shot experiment like that. And so for decades, prediction became the 'P-word' and nobody used that word at all. But now, science is marching on. And we're seeing things like ETS events, we're seeing things like turbidite evidence for clustering [of Cascadia's earthquakes over time], and the [possible triggering] relationship to the San Andreas. People around the world are seeing other, similar kinds of relationships. And while it's far from prediction, it's progress."

"It may be that we never get to the 'You're gonna have an earthquake *next Thursday*' sort of scenario," added Goldfinger. "But I think it's entirely possible that we'll get to a point where we can say, 'Sometime in the *next decade* we're highly likely to have something happen.' And I think that sort of thing is on the horizon, in the not-too-distant future."

Lori Dengler at Humboldt State, who began her career as a prediction optimist working initially with William Bakun and Allan Lindh on the Parkfield experiment, eventually lost her enthusiasm for trying to read seismic tea leaves. In her opinion building stronger buildings and making communities more resilient should be the higher priorities. As for the current ability to forecast Cascadia's inevitable failure, she said, "Well, I'll tell you something I'm absolutely sure about. The next Cascadia earthquake is one day closer today than it was yesterday."

Put the question a different way and you face another quandary. What if we *did* achieve a breakthrough in the science? What if the experts had another success like Haicheng and then became courageous, or foolhardy, enough to issue a prediction for Cascadia or the San Andreas? Would politicians and public officials know what to do with the information? Imagine you are the mayor, the police chief, the premier of British Columbia or the governor of Washington, Oregon, or California and the scientists come to your office early one morning and say, pretty much as Cao Xianqing did, "We think a major earthquake will happen today or this evening." How would you respond? What would you do?

"Some people would question that if you have a prediction—if it's not accurate enough—that you may cause more disturbance to life than to save life," Kelin Wang said. "If you predicted something that didn't happen—if you shut the factories and people moved away—and nothing happened? It's a complicated issue. It's both scientific and social." He shook his head and smiled. "It's very complicated."

PART 3 SHOCKWAVES

CHAPTER 21

Facing Reality: Cascadia Equals Sumatra

Vasily Titov's flight to Chicago was canceled at the last minute, so he was destined to spend Christmas alone in Seattle without his wife. She had taken an earlier flight and was already back east visiting relatives when unexplained airline woes at Sea-Tac Airport ruined Titov's holiday in December 2004. "It was a sad moment for me that I had to spend Christmas day and Christmas night by myself," Titov said, quietly mocking himself. "I had nothing better to do than go to the office and play with my model," his computer model of a large tsunami. Late that afternoon he took the scenic route along Sand Point Way to NOAA's Pacific Marine Environmental Laboratory on Seattle's Lake Washington and fired up the hard drive.

As the evening dragged on he tinkered with numerical codes he had developed to simulate the behavior of the massive waves that had smashed across Okushiri Island off the coast of Japan back in 1993 and was only half listening to CNN in the background. Suddenly a news bulletin caught his attention. There had been a seismic shock in the Indian Ocean. "The earthquake was small at first," Titov recalled.

"But it got bigger in front of my eyes." Later reports said the magnitude could be 8 or even higher.

Then came news that a tsunami had been generated off the coast of Sumatra and that people had been killed as far away as India. Titov was now riveted to the screen. "I saw the first reports of the deaths from the tsunami—the reports from CNN that fifteen people died in India. In *India*," he repeated, astonished. "It was thousands of kilometers away from the source. So that definitely gave me some idea that it's a huge event." By now it was late night in North America—Christmas night—and December 26 in Sumatra and India. Titov relaunched his model and started looking for data from the Indian Ocean that might allow him to re-create the tsunami while it was still happening.

Titov's boss, Eddie Bernard, remembered the first call in the wee hours of the morning. "I was in bed and the phone rang and ABC News said, 'Could you tell us about the tsunami?' And I said, 'Uh, I'll have to go check my internet and email messages and I'll get back to you.'" By the time Bernard joined Titov at the lab, the picture had changed radically for the worse.

"Vasily went to work on the model and Shirley, my wife, and I answered the phones," said Bernard. "The telephone calls were coming in, many more phone calls than we could ever address," Bernard confessed. "I mean I was doing interviews in Australia, radio stations in India—all over the world—London. And we were trying to provide some graphical information to the broadcast media at the same time we were trying to educate people about what was actually happening. Because, you see, in the Indian Ocean they had never seen or experienced anything like this and many people didn't even know what the word *tsunami* meant."

Titov, meantime, was hoping the codes he'd used for Japan and the North Pacific would work just as well for the coast of Sumatra, where the tsunami had been generated. Fortunately some research had been done relatively recently on the offshore subduction zones in that area and bathymetric grids were available for much of the Indian Ocean. There was very little information, however, about how big the first wave had been.

"The only data we had available was a tide gauge record on Cocos Island," Eddie Bernard explained, "which is south of the source in the Indian Ocean." Titov used the reading from the Cocos Island gauge as a proxy for a deep-ocean gauge, estimated the tsunami height, and punched the number into his computer as a starting point for the model to begin running a simulation. "Vasily was able to take that information and invert it into his models and get a fairly accurate representation of the tsunami," said Bernard.

Titov's math turned out to be pretty much bang on. When they ran the simulation it revealed things about wave behavior that no one had seen before. It showed, for example, that a tsunami could bounce off one island and hit the back side of another with even greater force.

Titov, standing in front of a floor-to-ceiling map of the world's oceans spread across the wall of a large conference room wall at PMEL, pointed to Sri Lanka, thousands of miles northwest of Sumatra. "So it came to Sri Lanka and hit hard." Titov's finger traced the leading edge of the tsunami to the southeastern beaches of Sri Lanka, where the first pulses crashed ashore with deadly force.

"On the back side, it was protected," Titov continued, pointing to the lee side of the island between India and Sri Lanka. "It was sort of shielded from the first wave. But then the wave bounced off the Maldives," his finger followed the path across to the next neighboring island chain, southwest of Sri Lanka. "It reflected from here [the Maldives] and then hit the backside of Sri Lanka with much *stronger* waves."

And there was stunning home video footage to prove the point. A mound of water slammed against the seawall at a luxury resort, shot a geyser of white spume into the sky, and then surged across the pool deck, sweeping away everything in its path. On the *back* side of Sri Lanka.

Titov's model also illustrated how the train of monstrous swells would turn corners around continents and eventually hit beaches on the opposite side of the planet. Big mountain ranges at the bottom of the Indian Ocean steered the on-rushing tsunami in new directions, according to Titov. He shifted his hand to the southern Indian Ocean and pointed to an undersea ridge. "You don't see them on the surface, but the wave does see them."

In a computer lab down the hall, he showed me how his model had replicated the motion. As the tsunami approached a ridge of very large undersea mountains, the wave began to bend and change trajectory. "The wave feels the shallow water and it slows down over this ridge," Titov explained. The tsunami scraped along the edge of the mountain range and friction slowed the left side of the wave down. The right side—still in deep water—continued to move at a higher speed. The difference in speeds from one edge to the other caused the wave to turn. In essence the undersea mountains became a wave guide, warping the deadly swells in a new direction.

But only up to a point. "If the mountain range turns sharply," Titov continued, "the wave would not turn. It will leave the guide." That's exactly what happened at the bottom of Africa. The tsunami veered off a mid-ocean ridge and rebounded like an eight ball to the corner pocket—around the tip of the African continent, crossing from the Indian Ocean into the South Atlantic. An example of "very interesting physics," Titov said, with barely controlled enthusiasm.

Roaring past the Cape of Good Hope, the sea monster moved up and across the Atlantic, coming ashore and leaving footprints on the beaches of South America. "A tsunami generated in the Indian Ocean— in *Sumatra*, half the world around—turned out to be a meter in *Brazil*," Bernard pointed out, with a shrug of amazement. "Fortunately it was at low tide, so it didn't do any damage." Sumatra's biggest wave even showed up on tide gauges as far away as Halifax, Nova Scotia.

The computer model predicted that the tsunami would circle the entire planet, and physical data from beaches and harbors around the globe confirmed the prediction. Then a bit of luck added another layer of confirmation. Two weeks after the tsunami, scientists at NASA notified Eddie Bernard and Vasily Titov that one of their satellites just happened to be overhead precisely when the tsunami was crossing the Indian Ocean.

"This tsunami was big enough in the open ocean—about forty centimeters—that [the satellite] actually detected it," said Bernard. Forty centimeters (15 inches) didn't sound big to me until Bernard reminded me that most of the wave was beneath the surface, reaching all the way to the bottom. The ocean was several miles deep and yet the wave still lifted the entire surface of the sea another forty centimeters. From that perspective it was an enormous mountain of rolling water. "And that is exactly the kind of data that we needed to verify our model," Titov said.

"It was an incredible match," Bernard continued. "So we were extremely happy." He grinned and then added, "Vasily was of course beside himself because this was the first time we had seen an open ocean tsunami of this size. And to model it correctly was quite satisfying."

While it might sound callous to be satisfied about successfully creating a digital clone of a killer tsunami in a computer, what Titov and Bernard and dozens of other researchers around the world learned from the Indian Ocean will no doubt save lives the *next* time this happens. When it does, detection systems now in development will be tripped as soon as the tsunami begins to move. With the knowledge gained from putting Titov's model to the test, it should be possible to say with some degree of certainty how that next monster will behave, what communities will be at risk, and how far up the beach the waves are likely to reach.

As Eddie Bernard put it, "Scientifically we have a wealth of new information—I mean, unprecedented information—that will guide us and improve everything we do in this whole field. But socially I think the most important thing that's happened is it's raised the awareness of tsunamis through the whole world. People take tsunamis seriously now."

To underline his point, Bernard told us about the first real-time application of NOAA's updated tsunami warning program in the aftermath of Sumatra. On November 15, 2006, a Kuril Island earthquake in the North Pacific generated a wave train big enough to create damage over a long distance. The swells began moving east across the Pacific, tripping the alarms on the deep-ocean warning buoys built and deployed by NOAA for just this purpose.

When one of the pressure sensors anchored on the bottom of the Pacific detected the extra weight of a larger than normal wave passing overhead, it transmitted a signal to warning centers in Alaska and in Hawaii. Now, instead of knowing only that an undersea earthquake had occurred, NOAA personnel knew for certain that swells of a potentially dangerous size had been generated.

Computers immediately ingested the data and spat out a prediction about where the waves would go and how big they would be when they got there. As the leading edge of the tsunami pounded across the Pacific from the Kuril Islands toward Crescent City, California, the computer produced a forecast. "There were ten waves," said Bernard, "and the model picked this up and actually replicated it *before* the tsunami arrived in Crescent City. It predicted that number nine wave would be the biggest. And guess what—number nine wave *was* the biggest."

Although none of them was as large or as vicious as the killer from Sumatra, the power of the moving water stunned those who saw it firsthand. Grady Harris, a grizzled and weather-beaten fisherman we met on the docks at Crescent City, told us he had heard the warning from NOAA and desperately tried to get his fishing boat out of the harbor and into the relative safety of deeper water. He made it just outside the breakwater, then got caught in a twisting torrent of seawater.

"In forty years on the ocean I've never seen that kind of a situation before," said Harris. "It was like trying to drive the boat in a washing machine. It just turned the boat sideways, turned it—spinning it around." He shook his head and stared across the dock, reliving the moment. "The awesome power of the water . . ."

Harbormaster Richard Young saw the surge boiling through the harbor entrance. He and all the others standing on the docks that day were mesmerized. "Water came in so fast on that larger wave that it actually went over the top of the concrete floats."

"It broke up the docks and scattered boats all over the harbor," said Harris. "There was several sailboats sittin' up on the dock from just the force of the wave."

To put this relatively small tsunami into context, the largest surge was only six feet high (1.8 m) from peak to trough. Not even as high as some of the regular storm tides in Crescent City harbor. The difference, according to Richard Young, was the speed—the manic rush—of the incoming waves. "The fact is that we have six-foot and eight-foot and ten-foot tidal changes here all the time with no damage," he said. "The difference is that they [the tides] happen in six hours instead of ten minutes." The damage estimate for this small, non-fatal tsunami was nearly $10 million.

On the positive side, NOAA's ocean warning buoys and the new tsunami models, refined and upgraded by the Indian Ocean experience, had given scientists reason to believe they'd made a major breakthrough. Now when an undersea earthquake sets off seismic alarms, there will be information about wave generation as well. Emergency planners like Stephanie Fritts and Sheriff Benning won't have to play guessing games with nature, wondering whether or not to issue evacuation orders to the citizens of Pacific County.

Vasily Titov extended the thought. "We cannot say when the next big earthquake is going to happen, but from the moment a tsunami is generated our models can actually tell you pretty well what happens next. How high the tsunami wave is going to be at the coastline, how big the impact is going to be at a particular location. The only thing we have to know for that is the actual measurement of the tsunami wave."

The success of the tsunami models, even though they're "never going to be perfect," he said, "makes you feel that—gosh, all this mathematics that you learned in high school and the math at university can actually pay off and save lives. That's a pretty amazing feeling. Dry mathematics applied correctly—it can save lives."

Chris Goldfinger, the marine geologist from Oregon State University, agreed with Titov. While still at sea off the coast of Sumatra, collecting mud-core samples in order to study how this earthquake happened, he couldn't avoid thinking of home. With a mud core sliced open on his laboratory workbench and the sea gently heaving beneath the hull of the *Roger Revelle,* Goldfinger drew the connection.

"I have to admit mud is not very exciting to look at. It just looks like some sand and some mud. But now every time I look at these cores, I see that giant breaking wave . . . It sort of brings home what these things really are," he said. "Even though we work in the theory of all this, to actually *see* that for real was just stunning, and horrifying . . . I think everyone's pretty mindful of the reason we're here," he continued, with members of the Sumatra science team looking on, "and that maybe some of this research might help in some way."

He paused for a moment and then tied the two stories together. "The same thing applies to Cascadia. I live there. And every time I drive to the coast, I see towns that are not long from now going to be under water from the next tsunami . . . The Cascadia Subduction Zone earthquake and the tsunami that'll come with it will be virtually identical to the one in 2004 in Sumatra. It'll dwarf 1906 [in San Francisco]. And Katrina. It'll be many *dozens* of Katrinas all at once. Coastal towns from northern California to Canada will be virtually wiped out. And there'll be significant damage in all the coastal *cities* along there as well—all at the same time."

Goldfinger agreed that knowing what Cascadia's fault is capable of had utterly changed his perspective about living on the edge of a continent. "It's a little hard to go to the beach and just hang out there and enjoy it."

Garry Rogers at the Geological Survey of Canada told me that what happened in Sumatra should have come as no surprise—and yet it did. "To all the scientists it was obvious that that's the kind of thing that happens," Rogers said, matter-of-factly. "It was perhaps more severe in terms of death toll than most of us would have thought. But what it did, I think, to the general public—and what we tried to translate to the general public—is that *Sumatra is Cascadia*. Those zones are the same size; what happened in Sumatra is what happened in Cascadia in 1700, and many times before that."

Now Rogers' voice was rising. "That's what we're talking about, *big* shaking. It damages a *lot* of buildings. And then a tsunami comes in on the shore. And we need to be able as a society to deal with that situation."

CHAPTER 22

The Next Wave: Thinking the Unthinkable

To me the good news is that people living on the edge of North America are finally beginning to respond. Outport communities on the west coast are taking advantage of new flood maps that show how high up the beach Cascadia's wave is likely to reach, what parts of town will be inundated, and where the safe zones on high ground should be located. Tsunami evacuation routes have been mapped and signs posted. People are going to night classes to learn what they need to know in order to survive. Evacuation drills are being held. And in some cases schools, hospitals, and other vital public buildings are being reinforced or relocated.

Scientists from at least half a dozen universities in the United States and Canada are creating or updating numerical models that use detailed bathymetric maps of the local harbors and offshore terrain to predict much more precisely how far uphill and inland the turbulent muck is likely to travel. They're doing on a local level what Vasily Titov's model did on a global scale. Computer simulations of Cascadia's tsunami have been generated for the city of Victoria and for the fishing village of Ucluelet on the west coast of Vancouver Island, for Cannon Beach and Seaside on the Oregon coast, and dozens of others are in development.

Not every community in harm's way has a computer model to map the inundation zone. At least not yet. Some towns such as Port Alberni, which was hammered by the Alaska tsunami of 1964, don't have detailed bathymetric grids to work with because hydrographic survey ships are expensive to operate and government research budgets have been slashed. Rather than wait for public funding to catch up with grim reality, Port Alberni's emergency planning team took matters into their own hands—literally.

Knowing from experience how waves coming in from the coast can get amplified by the narrow canyon walls of a fjord like the Alberni Inlet, local planners asked the experts at the Pacific Geoscience Centre for their best guess about the height of a seawater pulse coming from Cascadia's fault. Then they took out a standard topographic map of the downtown core along with a red marking pen.

Bob Harper, the head of emergency planning for the city of Port Alberni at the time, walked me through the exercise. "The best advice that we've received so far from the scientists is—because of the funneling effect in the Alberni Inlet—that we can expect somewhere in the twenty-meter range of inundation," said Harper. "So that's twenty meters high . . . Not twenty meters in, but twenty meters up."

"That means a sixty-foot wall of water?" I asked, trying to imagine the downtown waterfront.

"Yes," he said quietly. A technician working with Harper began tracing the contour in bright red ink across the heart of downtown. "There's a band of residences along the riverside here," his hand swept across the map following the contour around the harbor. It was clear that most of the central business district, all of the industrial waterfront, the pulp, paper, and lumber mills, would be inundated.

Cascadia's wave would be larger by far than anything seen in 1964, mainly because this subduction zone—birthplace of the tsunami—is so much closer to home. From the moment the ground begins to shake, places like Ucluelet and Tofino on the west side of Vancouver Island, along with Cannon Beach and Seaside on the Oregon shore and the beaches of Pacific County, Washington, may have as little as fifteen minutes before the first wave makes landfall, with as many as eight or ten more en route behind it. And so the obvious question arises: what could or should a person standing on the beach do to survive? The logical answer is to head for higher ground. And do it right away.

I gave my stopwatch to Patrick Corcoran, who strapped it on his wrist and got ready to run. He stood only a few steps away from the wide, sandy beach at the base of the Lewis and Clark statue in the traffic circle at the western terminus of Broadway, the main drag in Seaside, Oregon. Young and fit, probably in his late thirties, Corcoran is a surfer by choice and an employee of Oregon State University by profession. As a "hazards outreach" specialist, he helps people along the coast plan for and come to terms with some of the realities of life on the edge— things like major winter storms, and getting ready for "the Big One" from Cascadia's fault.

Corcoran is demonstrating the official evacuation route, which begins at the beach promenade and follows Broadway, the main east–west business corridor, all the way across the downtown core to higher ground on the east side of town. As I give Corcoran the nod, he clicks the stopwatch and starts running.

In a real emergency, running or walking would probably be the only way to get out of town fast. Previous experience with tsunami false alarms in Seaside had already taught local residents that vehicle traffic hits gridlock almost immediately. It was a busy summer day so the sidewalk was crowded with shoppers and tourists out for a stroll. Corcoran jogged at a brisk pace, zigzagging through the throng.

The Pacific coastline here runs almost exactly north–south, so to get away from the ocean on the west side you have to move toward the first rank of low hills in the Coast Range mountains on the east. The city of Seaside is built on the wide, flat delta of the Necanicum River, so anyone trying to outrun a tsunami would have to hustle to get across the two bridges that span branches of the river, hoping the earthquake and the incoming tsunami had not already knocked the concrete decks off their pilings.

By the time Corcoran got past the river and started uphill on a winding switchback road, he was breaking a sweat and breathing hard. He had covered more than twenty city blocks without gaining any altitude. Now the road started to climb steeply. He already knew from studying the map contours exactly which house he had to reach in order to get himself at least fifty feet (15 m) above sea level and presumably beyond the reach of the biggest waves likely to come from Cascadia's subduction zone.

When he crossed the imaginary finish line, he stopped and clicked the watch again. "So, seven minutes and thirty-three seconds," he huffed. "Not too bad, but it was a hard run." On a nice sunny day under ideal conditions, he certainly would have made it to high ground in plenty of time. But what if the earthquake happened on a stormy winter's night? Powerlines and trees would be down and all kinds of obstacles would be in the way.

And what about those not as young or physically fit as Patrick Corcoran? The likelihood that the vast majority of people could make that run before the first tsunami surge hit the beach seemed pretty slim to me. For those who hung around to *watch* the incoming waves, mesmerized by the spectacle as so many were on the beaches of Sumatra and Thailand, the chances of survival would be even less.

Corcoran has a set of simple guidelines he explains to anyone who will listen. Rule number one: if you're anywhere near the coast in a subduction zone and you feel the earth begin to shake, start moving to higher ground as soon as the shaking stops, or sooner if you can. Rule number two: if you're living in or visiting a coastal earthquake zone you should already know where the high, safe ground is and how to get there. Grab a map, study the evacuation routes, and always have a sense of where you are. Rule number three: don't wait for a warning siren because there probably won't be one. Your only warning will be the violent shaking of the ground, so don't wait for someone in authority to tell you to run. Rule number four: if you stand there long enough to *see* the incoming tsunami, there's almost no chance you'll outrun it.

These are the kinds of things Corcoran recites when he does his outreach sessions. "When I came here in 2003 for the Coastal Storms Program," he begins, "it was all about severe winter storms and flooding and those kind of more chronic or constantly occurring events. Then the 2004 Sumatran earthquake and tsunamis occurred, and while my *focus* was on storms, I had to ask myself as a coastal hazards agent, is this [tsunami threat] something that I should be paying attention to? And so I asked myself three questions: One, is it likely? Two, would it be bad? And three, can education do anything to improve the situation?"

In a heartbeat he answers his own questions: "Yes, yes, and Lord, I hope so. Yes it *is* likely. Yes it *would be* very bad. And education *will*, hopefully, go a long way towards improving the odds for most people."

The first thing to know, according to Patrick Corcoran, is that if you hear a warning siren you shouldn't panic. It probably means there has been a distant temblor in Japan, Alaska, or Chile. "So if you heard a siren, or understood a tsunami was coming, and you didn't feel the earthquake, Alaska is the closest place it's going to be and that's three and a half to four hours away," says Corcoran. "So the good news is— you have time before a small and not so bad tsunami comes."

If it's a seismic jolt closer to home, the Big One from Cascadia's fault, the sirens won't have time to go off before the first wave gets here. That's because NOAA's warning buoys are anchored farther out in the Pacific—beyond the Cascadia Subduction Zone—in order to provide plenty of warning time for those long-distance waves from Japan, Alaska, and elsewhere. Cascadia's fault, being much closer to the shoreline of North America, will send out waves both east- and westbound. And the eastbound waves will hit the beaches of California, Oregon,

Washington, and British Columbia *at the same time as* or even before the westbound waves hit the warning buoys and trip the alarm. Until this chink in the armor of the tsunami warning system is dealt with by some newer generation of detectors closer to shore, the only real warning anyone in Cascadia will have is the violent shaking of the earth.

Corcoran begins to distil the to-do list. The first thing to figure out is whether the quake and tsunami are from far away or closer to home. Did you feel the earth move? Or was it a siren you heard? The implications are obvious and the necessary responses completely different.

"The second thing you need to know," says Corcoran, "is where are the safer and less safe places. And not just where you live, but where you live, work, and play." In other words, plan an evacuation route to safe ground from any place where you spend a considerable amount of time.

"Develop an eye for the landscape," Corcoran suggests. "So, when the Big One occurs and we're out driving around conducting our lives, we'll have some sense of, 'Wow, I'm in a bad spot. I need to get over there.' Or, 'Wow, I'm in a good spot—relative to tsunamis—I'm gonna stay put.'

"The third thing," Corcoran continues, "is how to reconnect with your loved ones. After the Cascadia Subduction Zone earthquake and tsunami, it's going to be very difficult to get a hold of family members. There will be no phones. And you will not be driving anywhere." He suggests that families pick a rendezvous point somewhere on safe ground and plan for all family members to meet there. That way everybody knows that everybody else will eventually make their way to the same place even if there's no way to communicate.

Corcoran asks people to imagine the nightmare of successfully escaping the shockwave and then deciding to go back into the disaster zone in search of a family member who has already escaped and is en route to a rendezvous point or rescue center. A person could die for lack of planning. "Actually, the important thing is sitting around the kitchen table with your family and thinking through some scenarios. If *this* happens, what would *you* do? Well what if you're at school? What if you're at work?"

What it comes down to is this: when the Big One hits, you're on your own. This is all about self-reliance. And helping your neighbors.

What if, as some emergency planners have suggested, people were able to escape the inundation zone by climbing a set of stairs instead of running halfway across town horizontally? The concept of "vertical evacuation" seemed to make instant good sense. To compare the official evacuation route with a hypothetical plan B, Patrick Corcoran agreed to run for his life again.

Poised again at the base of the Lewis and Clark statue, this time he ran only three city blocks to the nearest easily accessible building that was more than three stories tall. The most recent computer model for Seaside has suggested the waves from Cascadia could be ten meters high, or a bit more than thirty feet, a monster by any measure. Just one block inland from the beach is a public parking building with four or five levels, just enough vertical amplitude to get us above a thirty-foot wave.

Corcoran didn't need to click the stopwatch this time because it was obvious he would make it to the parkade well before the imaginary tsunami hit the seawall. He took off at a brisk jog from the promenade to the first traffic light, where he hooked a quick right and headed for the stairwell door at the base of the building. The additional benefit of a parkade structure is the gently inclined ramps. People in wheelchairs or those who cannot thunder up the stairs as Corcoran did would still be able to gain some elevation without having to go all the way across town.

Every beach town I've ever seen has always had a shortage of parking spots. What if city hall—with help from senior levels of government—were to solve their parking problem and save lives at the same time? All they'd need would be a building designed and engineered to withstand both the seismic shaking and the torrent of water.

∞

It just so happens that FEMA, the Federal Emergency Management Agency in the United States, the USGS, NOAA, and all five Pacific Coast states have already commissioned a study of that very idea. Not a parkade, necessarily, but earthquake- and tsunami-resistant vertical evacuation shelters. The engineering study was only one component of the National Tsunami Hazard Mitigation Program created by the U.S. Congress in October 1996—the product of lobbying efforts by people like Eddie Bernard and Lori Dengler in California in the wake of the Petrolia earthquake.

From deep-ocean warning buoys and computer models to estimate and predict tsunami run-up and inundation zones town by town and beach by beach, the United States, at least, seems ready to take seriously the job of making coastal communities "tsunami ready." Harry Yeh, a civil engineer at Oregon State University and one of the three principal investigators on the shelter study, believes most of the critical engineering problems could be solved and the proof was in Sumatra.

In 2005, as engineers studied the tsunami aftermath in Indonesia and Thailand, everywhere they looked, "well-engineered, reinforced concrete structures were still standing," said Yeh. He showed me a picture, drawing my attention to an apartment block or hotel right at the waterline in Banda Aceh. "Even though the structure was completely inundated to the roofline," he said, "the structure itself is still standing. So our experience says that if you have a well-engineered concrete structure, I think those can be used for tsunami shelters."

Yeh also showed me other pictures of an odd-looking, cone-shaped building erected in a coastal town in Japan, where the concept of vertical evacuation has been studied, debated, and implemented already. In some places the top floors of apartment blocks, warehouses, and public buildings have been designated and prominently marked as tsunami shelters. Stairwells and doors to the rooftops are never locked. Local residents have been assigned specific numbered or marked spots for their families in case of an emergency. Regular drills are conducted in which able-bodied neighbors practice carrying senior citizens and disabled people to the top floors.

In the small town of Taiki, the Nishiki Tower was custom built to survive the effects of the expected Tokai earthquake. It was also hydrodynamically designed to withstand the forces of fast-moving water. With rounded, conical walls and a spiral stairway to the top, it has shelter rooms and emergency supplies on the upper floors. The thing is—it *looks* odd—like a tall, white lighthouse in the middle of town, completely out of place. And that causes out-of-town visitors to stop and ask questions.

"If I see such a tower," Harry Yeh speculated, putting himself in a visitor's shoes, "I'm gonna ask the people, 'What is this?' So everybody will know that's a tsunami shelter." He smiled. In essence, looking odd or out of place could help a tsunami shelter save lives. "I think this is a very important component of the design," he said. In the meantime he and a study team continued to work on a new set of building code guidelines for vertical evacuation shelters.

Among the engineering challenges, according to a report issued at the end of the first phase of the study, was that designing a building to withstand a seismic shock is in some ways the opposite of what you'd need to survive a tsunami. To ride out an earthquake, a building needs "flexibility, ductility and redundancy." To outlast a tsunami it needs "considerable strength and rigidity, particularly at the lower levels." But Harry Yeh insisted these requirements "need not be contradictory" and stressed that both had to be taken into account.

The foundations of a tsunami shelter would have to withstand not just the violent shaking but the soil liquefaction that often accompanies a quake. They must be deep enough below unstable soil to be anchored on firm bedrock. The building itself would have to provide enough floor space for evacuees and be tall enough to stand above the largest expected wave. The walls would have to be strong enough to withstand the battering-ram effect of water-borne missiles (floating cars, logs, lumber, and other debris). It would have to be fire resistant since quakes and tsunamis always cause numerous fires to break out. The final design requirement would be resistance to scour. The foundations of a shelter would have to withstand the rapid rise and fall of fast-moving water that would "loosen the soil skeleton" around the building, possibly causing collapse.

While the challenge sounds daunting, the report underlines the obvious concern that vertical evacuation may be "the only choice for human survival" in many coastal communities. Because of the engineering complexity shelter designs will probably have to be done on a case-by-case basis. Every beach and the bottom of every bay is a little bit different.

CHAPTER 23
Watching It Happen, Wishing It Wouldn't

Harry Yeh, Patrick Corcoran, and Chris Goldfinger met on the campus of Oregon State University in Corvallis for one of the most riveting demonstrations of the power of moving water I'd ever seen. Behind the blue-gray corrugated metal walls of a hangarlike building that looked big enough to hide a blimp, in a wave research basin half the size of a football field, researchers led by civil engineer Dan Cox had built a scale model of the town of Seaside, Oregon. The object of the exercise was to test the effects of a tsunami from Cascadia's fault on a detailed physical replica of Seaside's downtown core. Computer models had already predicted what would happen, but how would real water behave compared to a hypothetical digital clone?

Graduate students and technicians from OSU had spent months building plywood surrogates for each of the main beachfront hotels, commercial buildings, parkades, and homes in the downtown area. They built an inclined platform and poured a concrete floor at exactly the same angle as the sea floor and beach. They constructed a breakwater exactly like the real one that stood beneath Seaside's popular promenade. They marked out a duplicate street grid and used bolts and nail guns to anchor all the buildings into the concrete. From above it looked remarkably like the real thing, only fifty times smaller.

Dan Cox and his team then programmed a sophisticated set of computer-controlled mechanical paddles at the far end of the basin. The system was capable of generating a scaled-down version of Cascadia's wave: one-fiftieth the size of the real one oceanographers and marine geologists expect to see crossing Seaside beach some day in the unpredictable future.

A special-effects camera team filmed the experiment (for the *Shock-Wave* documentary) so others could observe the results. To visually slow down a lump of water moving fifty times faster in the tank than the real wave would sweep across the beach at Seaside, we used a special high-speed camera that could shoot up to 1,500 frames per second and still deliver a high-definition color picture. We used a snorkel attachment to create a pedestrian's eye view of the tsunami as it moved up Broadway. We were able to play back the wave experiments on a large-screen, flat-panel TV display. On a work table beside the giant monitor, a computer terminal had been set up by Patrick Lynett, a scientist from Texas A&M University who had been working for months on a parallel experiment to refine a numerical model designed to match the bathymetry and layout of Dan Cox's model of Seaside. They would run their waves simultaneously and compare results.

For Lynett and the many others involved in the computer modeling of tsunamis, the running of a wet physical replica of Cascadia's wave in a test basin like this at OSU would provide a crucial benchmark—a reality check for the mathematics. If the two models showed pretty much the same results, then an extra measure of confidence would be gained for the computer simulations. A large physical replication in concrete and plywood for each of the dozens of communities threatened by Cascadia would never be affordable, either in dollars and cents or in the amount of time it would take. But if a computer model could reliably tell you the same thing, physical models wouldn't be necessary.

In principle, if Lynett's model worked well for Seaside, then it could be reprogrammed and modified with new bathymetric and street grid details for the next town on the coast, and a more realistic appraisal of the inundation zone and specific levels of risk could be had much sooner and at lower cost. At least that was the theory and the reason that people like Dan Cox and Patrick Lynett were eager to see what happened next.

Chris Goldfinger, back from his research cruise to Sumatra, offered a sobering caution. The numerical simulation of anything as sloppy as moving water is extremely difficult to do. It was hard enough to work on a broad, oceanwide scale as Vasily Titov had done, but even more challenging when you tried to zoom in to detailed street grids and individual buildings in a single town. The tighter the grid, the more exacting the model, the greater the chances for error.

"The best computer models now are working hard at quantifying the flow [of water] around *one or two* objects," Goldfinger explained, "a cylinder, a bridge piling, something like that—a relatively simple case—just because the computational time is enormous." When the myriad three-dimensional obstacles in a real harbor and town are assigned numerical values—the friction coefficient for water moving over the sandy ocean bottom, a different level of friction and drag once the swell crests, crashes over the seawall, and begins moving over dry ground cluttered with buildings, cars, trucks, trees, and lumpy terrain—it gets a lot more difficult.

After a quick check by portable radio with the crew standing by in the control room to confirm that the computer and the paddles were ready, Cox turned to his visitors. "So what you're going to see next," he explained, "is the rough equivalent of the five-hundred-year-event." Meaning the full-margin rupture of Cascadia's fault that takes place on average every five hundred years. "So this is a twenty-centimeter lab scale or [the equivalent of] a ten-meter full-scale tsunami that is coming into Seaside." Quickly doing the conversion in my head, I tried to picture a surge of water more than thirty feet above the high tide, thundering toward the beach.

Cox gave the order and moments later the long row of paddles at the far end of the tank thrust forward at the calculated angle and speed. A dark swell began moving, quietly hissing toward Seaside. The wave crashed against the breakwater and shot a slice of foamy spume straight up. In the next heartbeat the on-rushing tide poured across the promenade and churned up Broadway, sweeping up toy school buses, cars, and trucks in a frothy vortex that quickly swamped the entire model.

As a matter of interest I noticed that one prominent multistory hotel right on the waterfront had been completely overtopped by the wave. The upper deck of the parkade building directly behind the beachfront condominium complex had remained dry. Patrick Corcoran, standing elbow to elbow with the other observers on the pool deck, noticed it too. Anyone who could get to the top of that building would probably have survived, but anyone stranded at street level or anyone trying to escape in a car probably wouldn't have had a chance.

When the sloshing finally stopped nobody said a word. Like Corcoran, many of those in the crowd knew the streets of Seaside well. It seemed as though everyone in the room was momentarily stunned, trying to absorb the news that most of the downtown area would be inundated by the pulses of seawater from Cascadia's fault.

Patrick Lynett, meantime, had called up a file from his computer model and was explaining how he'd created a numerical duplicate of the objects in the basin. The first frame of his simulation of Cascadia's wave looked like an animated cartoon of the physical model, a 3D aerial view of the miniature plywood buildings nailed to the concrete floor in front of us. His intention had been to make the layout of the numerical model resemble the built world of the research tank as closely as possible so that when he hit the run button, his computer-generated wave would face exactly the same obstacles. With any luck the digital tsunami would match the behavior of the real wave we had just witnessed.

As I watched the simulation play on the monitor, it looked pretty convincing to me. After pounding across the seawall, a dozen or more jets of dark-blue liquid surged past the first rank of buildings, pushing straight up Broadway and all the eastbound streets simultaneously, turning corners and twisting together like braided hair as they seemed to amplify themselves in thicker, darker currents along some of the narrow side streets.

"What we're looking at here is momentum flux," said Lynett, "which is a very good measure for the potential force of fluid for the tsunami as it comes in and inundates Seaside ..."

It wasn't clear to me that anyone really heard what he'd said. They couldn't take their eyes off the screen. This time we could see more clearly what the wave was doing because the computer had slowed it down to something resembling real time. With a click of the mouse, Lynett slowed it down even more and then stopped the action completely.

"If you look at this little building right here," he said, hitting pause and pointing to what looked like a small house on a side street several blocks away from the beach, "what actually happens is—you get a wake off of this building." He clicked play again and advanced the animation a few more frames. A thick wedge of the inky blue water pouring east on one street ricocheted off a larger commercial building across the street from the house. "It bounces off the side of this building—and additionally you get a large wake off of *this* building." Now we could see the almost synchronous arrival of another tongue of water on the next parallel street. "And those two line up and just *pound* this tiny little building," said Lynett.

Like pool balls bouncing off padded rails, like Titov's wave bouncing off the Maldives to hit the back side of Sri Lanka, these two strands of liquid energy careened off two multistory commercial buildings in the digital clone of Seaside, curled together in the middle of a side street, and combined forces. The computer animation changed the color of the flow to bright yellow and red to indicate the magnified amplitude and intensity.

"And so what you'll see," said Lynett, "if you look at it," and everybody clearly was glued to the screen, "you'll see that dash of red shoot here and then bounce. And then—just eyeball right on this building." He paused as if there was nothing else to say, then added quietly, "It would be extremely damaging . . ."

It would take months of detailed side-by-side comparison to see how closely the computer simulation and the wave-tank model had matched up, but they looked remarkably similar to the untrained eyes of those of us standing poolside that day in Corvallis. In the end some combination of the two approaches would probably emerge from this experiment to create a refined and updated system for predicting the effects of tsunami attacks on other coastal communities. "You sort of have to boot-strap between a physical model in a wave tank and a computer model to validate one against the other for the things that you can test," Chris Goldfinger explained, "and then go beyond the capabilities of either one by using them together."

After watching Lynett's computer simulation, it was time to play back our HD video. Because the physical model of Seaside had been built on a scale of one-to-fifty, we needed to slow the frame rate of our pictures by the same ratio—and the high-speed camera allowed us to do that—producing a slow-motion image that looked almost identical to those tragic home videos from the Indian Ocean. Now we could see the swelling mound of water as it hurtled toward the surrogate Seaside. At some critical point along the beach where the ocean bottom angled upward, the leading edge slowed down long enough for the back of the wave to catch up with the front. The swell piled up like a surfer's dream, curled forward, and then broke under the force of gravity in a hissing bore of fast-moving water.

As it shot across the last stretch of beach toward the base of the promenade a knife-edged geyser of spraylike jets, as if from a thousand vertical fire hoses, rocketed straight up in a perfect replay of the Sumatra waves hitting the wall of a resort in Phuket, Thailand. And just as mesmerized as those poor souls who stood like backlit deer at the foot of the palms waiting to see what would happen next, we observers at the OSU tsunami basin could not take our eyes off the screen.

Having crashed over the promenade, the wave continued pounding straight at us. The street-level view from the snorkel lens revealed the Broadway canyon between the seven-story condominium complex and the beachfront hotel across the street as the roiling water lifted a toy school bus, an ambulance, and several other vehicles and swept them away—exactly the way the tsunami did in Banda Aceh. A floating garbage truck crossed the sidewalk and slammed backward like a levitated battering ram into a two-story commercial building. A few frames of video later, the wall of tumbling junk rolled right over us and the picture went dark.

"Very sobering," said Doug Barker of the Seaside Fire Department when the video finally stopped. "It was . . ." he paused, searching for the right words. "It was actually a shock. I was—it took me more by surprise than I thought it would to watch the water roll through, between those buildings. And cascade *over* the buildings. So, yeah, it was very eye opening, even though I've been dealing with it for a number of years."

Another of the invited observers, Barbara Lence, a civil engineer from the University of British Columbia who had just completed work on a computer model of Cascadia's wave showing how it was likely to inundate much of the village of Ucluelet on Vancouver Island, was also stunned by the video. "One of the things that I take away is that this decision about whether we stay—or go—during one of these events is a very critical decision," said Lence. "Do we shelter in place? Or do we focus on emergency exit?" She was also aware as never before of "the importance of debris—vehicles and so on—floating in water, hitting buildings, hitting structures that may be providing safety."

"Hopefully we'll be able to use that videotape or the simulations to maybe wake up some folks in the local area," said Barker, "to show them—give them an idea of what to expect."

"If we could communicate that intensity to everyone," added Lence, "we might have a better chance at being prepared in these emergencies."

Wave-tank models and digital tsunamis were not the only kinds of experiments being conducted to anticipate the effects of Cascadia's next violent outburst. Coastal inundation zones are clearly not the only concern. By the spring of 2009, a new question had arisen: what would happen to the urban cores of major cities from Victoria and Vancouver to Seattle and Portland? These four cities are built on land that lies well to the east of the main fault, and some experts had suggested the rupture zone was far enough away that damage in the urban areas might not be as catastrophic as first thought.

When Roy Hyndman and Kelin Wang published their study of the locked part of the subduction zone in 1995, they calculated that the stuck and truly dangerous area of the fault lay nine miles (15 km) underground to the west of Vancouver Island and the beaches of Washington, Oregon, and California. The "landward limit" or leading edge of the locked zone extended "little if at all beneath the coast," which "limits the ground motion from great subduction earthquakes at the larger Cascadia cities that lie one- to two hundred kilometres inland."

With this in mind many emergency planners in the Pacific Northwest have worked on the assumption that Cascadia's magnitude 9 event would *not* be the worst scenario. Any of the much shallower faults in the continental crust could generate a magnitude 6 or 7 rupture directly beneath or very near an urban area, and this might actually cause more severe, localized damage. But in 2009 Timothy Melbourne and his colleagues suggested the locked zone was considerably nearer to the big cities, perhaps within 50 miles (80 km) of Seattle, for example. And if that's true, then we're right back where we started with Mexico City in 1985: what do we do about heavy shockwaves hitting high-rise buildings?

From a civil engineer's point of view, the elephant in the room has always been how an urban forest of tall towers, long bridges, freeway overpasses, and hydro dams would respond to the much longer *duration* of low-frequency seismic shocks from Cascadia's fault. The way it was explained to me, going from a magnitude 7 to a magnitude 9 means the intensity of the shaking doesn't change so much as the length of time it lasts.

Instead of forty-five seconds or a minute of shaking in a magnitude 6 or 7, the seismic shockwaves from a magnitude 9 could go on for four or five minutes. Like they did in Alaska. And nobody really knew how well tall buildings would stand up to that much horizontal motion, slamming side to side, undulating like metronomes fifty or a hundred times, flexing every beam of steel, stressing every welded joint, every slab of concrete, every pane of glass to the limits of endurance. All anyone could do was speculate about the outcome because there has never been a magnitude 9 in a city full of skyscrapers. Neither the Chile quake of 1960 nor the Alaska rupture of 1964, nor even the Sumatra disaster of 2004, shook a large, modern city with high-rise towers. Mexico City gave us only a hint of what might happen.

To oversimplify things just a bit, building codes in North America generally require engineers to design tall structures in earthquake country to survive the forces imparted by temblors up to magnitude 7. Specifications for a magnitude 9 don't exist because there have been so few of these megathrust events—only four in the last century—and so few measurements of the ground motion in real-world circumstances that nobody has a good grasp of how strong the vibrations would be when they hit any given high-rise. The cities of modern civilization have been built taller and taller with not a single full-scale test of what might happen in a megathrust temblor.

That was the problem Tom Heaton, who heads the Earthquake

Engineering Research Laboratory at Caltech, and Jing Yang, one of his doctoral students, decided to tackle next. Heaton and Yang built a computer model to simulate the effects of a magnitude 9.2 Cascadia rupture on downtown Seattle. They began with data from the 2004 Sumatran quake and Japan's magnitude 8.3 Tokachi-Oki rupture in 2003, along with geological data about various soil and rock conditions in and around Seattle.

They tested a series of hypothetical steel-frame buildings from six to twenty stories tall with older "brittle" and newer "perfect" welds at the critical joints. They ran the model several times, factoring in different possible distances from Seattle to the locked zone of Cascadia's fault, just in case the real rupture does extend farther inland toward the city. Their digital temblor made the ground shake for four minutes, the dominant shock being the low-frequency kind that caused so much grief and damage in Mexico City in 1985. The deep sedimentary soils in some areas of Puget Sound predictably amplified the waves, just as the dried-up lakebed did in Mexico City. In the new Caltech simulations, the soils increased the *duration* of shaking as well.

Heaton and Yang presented their results at the annual meeting of the Seismological Society of America in Monterey in April 2009 and journalists immediately wanted to know the bottom line. A glance at the poster Jing Yang had prepared for the meeting pretty much told the story: "Our simulations show that Seattle high-rise buildings with brittle welds have a significant potential for collapse."

When Yang started work on her numerical model, she was forced to simulate steel-frame buildings rather than large concrete structures simply because it was easier for a computer to predict how steel would bend and eventually fail. The fracturing of concrete was much more difficult to model. It was certainly plausible, according to Heaton, that older concrete buildings could be at even greater risk because they were probably even more brittle. But there was no reliable way to re-create that kind of failure in a computer.

Another reason to study steel was that the Northridge jolt of 1994 had shown scientists that brittle welds in older steel-frame buildings had failed more often than anyone had expected. Amendments to the building code made in the wake of Northridge have changed the way structural joints are welded, presumably giving newer buildings extra strength. And Yang's model did seem to confirm that newer towers would be stronger—but only up to a point.

If the rupture of Cascadia's fault happens to extend down below the west coast beaches to some point underneath the Olympic Mountains—closer to Seattle—the shaking would be much worse. And when that scenario was run in the Caltech simulation, *all* the high-rise buildings in Yang and Heaton's experimental model *collapsed*. Even those with "perfect welds."

Some of the science writers saw parallels to Mexico City and wanted to know more. "All the crummy little buildings that existed in Mexico City were completely undamaged," Heaton explained, offhandedly, to one reporter, "but the high-rise buildings, which were the pride of their construction industry, many of them collapsed. It wasn't just a matter of poor construction. It was a case of the wrong buildings being in the wrong place at the wrong time."

Low-rise, low-tech buildings simply did not vibrate or resonate at the same frequency as the big shockwaves generated by a subduction zone. High-rise buildings, on the other hand—even relatively new ones—constructed on thick sedimentary soils, vibrated *more* than any engineer or any building code had anticipated. They shook to the point of collapse. And what happened in Mexico would presumably happen in Seattle, Vancouver, Victoria, and Portland as well, according to Heaton and Yang's research.

"In general, high-rise buildings behave very differently from low-rise buildings," Heaton said. "They're primarily designed to be flexible. And in sharp, rapid shaking—during a *moderate*-size earthquake—high-rise buildings perform extremely well." But a magnitude 9 was quite obviously a different story. Yang told another reporter that there were approximately nine hundred high-rise towers within striking distance of Cascadia and half of those were built prior to 1994, when the new building code imposed tougher standards.

Reading between the lines, it was obvious to me that the number of tall buildings in danger would depend on how far "down-dip" Cascadia's fault slips when the Big One hits. If the locked part of the subduction zone—the part that will generate the shockwaves—extends farther inland than initially estimated, the impact on high-rise structures in big cities will be even more severe.

CHAPTER 24

Cascadia's Fault: Day of Reckoning

On a foggy spring morning just before sunrise, twenty-seven miles (43 km) northwest of Cape Mendocino, California, a pimple of rock roughly a dozen miles (19 km) below the ocean floor finally reaches its breaking point. On the same thrust fracture that rattled the towns of Petrolia, Ferndale, Eureka, and Arcata, two slabs of the earth's crust begin to slip and shudder and snap apart as Cascadia's fault finishes what it started back in 1992. That day could be only ten years away. Or two hundred years from now. Or it could happen tonight. And this is how I've imagined it will unfold.

The first jolt of stress coming out of the rocks sends a shockwave hurtling into northern California and southern Oregon like a thunderbolt—same as last time, only bigger. Ten times the magnitude and thirty-two times more energy. For a few stunned drivers on the back roads in the predawn gloom, the pulse of energy that tears through the ground looks dimly like a twenty-mile (30 km) wrinkle moving through a carpet of pastures and into thick stands of redwoods.

Telephone poles whip back and forth as if caught in a hurricane. Powerlines rip loose in a shower of blue and yellow sparks, falling to the ground where they writhe like snakes, snapping and biting. Lights go out and the telephone system goes down.

Cornices fall, brick walls crack, plate glass shatters. Pavement buckles, cars and trucks veer into the ditch and into each other. A bridge across the Eel River is jerked off its foundations, collapsing into turbulent eddies below, taking a busload of farm workers with it. A gasoline tank truck swerves to miss a car that's made a panic stop in the middle of Main Street. The tanker bounces over a curb, taking down a lamppost, crashing sideways into a corner store. Seconds later the wreckage explodes. The fire will be difficult to fight because water pipes under the street are broken. People are awake now and screaming, running dazed and wounded into the streets.

Seeing fragments of this happen through drifting shrouds of fog makes it hard for survivors to know how much is real, how much is their worst nightmare. With computers crashing and cell towers dropping offline, all of Humboldt and Del Norte Counties in California are instantly cut off from the outside world, so nobody beyond the immediate area knows how bad it is here or how widespread the damage. Same for southern Oregon. Despite the rising sun everything suddenly seems dark again. People living in the necklace of towns and villages along the coast are now officially on their own. No help will be coming any time soon.

On a spit of sand running down the western edge of Humboldt Bay, an air raid siren wails as residents in the former sawmill town of Samoa, barely a dozen feet (3.6 m) above sea level, bang through their doors—those that will still open, that are not twisted out of true by the violent shaking of their homes—and run, walk, or stumble as best they can toward slightly higher ground near the water tower. They know from past drills that the first wave could hit the beach as quickly as eight minutes from the moment of rupture.

At the USGS lab in Menlo Park seismometers peg the quake at magnitude 8.1 and the tsunami detection centers in Alaska and Hawaii begin waking up the alarm system with stand-by alerts all around the Pacific Rim. High-rise towers in Sacramento begin to sway. Early morning commuters emerging from a BART station in San Francisco feel the ground sway beneath their feet and immediately hit the sidewalk in a variety of awkward crouches, a familiar fear chilling their guts. Then another little rough spot on the bottom of the continent snaps off. The fault unzips some more.

Back in Petrolia, where the ground has been shaking for more than a minute already, the street now heaves like a trawler's greasy deck in a North Pacific gale. The entire Gorda plate has come unstuck. The outer edge of California snaps free like a steel spring in a juddering lurch—nine feet (2.7 m) to the west. The continental shelf heaves upward, lifting a mountain of seawater.

The new shockwave, from the latest broken rough spot, slams from the Gorda into the Juan de Fuca plate farther north—like gigantic train cars banging together—and thus the fault continues to rip all the way to Newport, Oregon, halfway up the state. The magnitude suddenly jumps to 8.6. A power surge blows a breaker somewhere east of town and feeds back through the system, throwing other breakers in a cascade of dominoes that quickly crashes the entire grid in Oregon, Washington, and parts of California, Idaho, and Nevada. A brownout begins in six more western states. The wireline phone systems crash in lockstep.

Then the asperity beneath Newport shears away. The fault unzips the rest of the way to Vancouver Island. The quake now pins seismic needles at magnitude 9.2. A pineapple express has delivered a long string of storms that are pelting rain from Cannon Beach all the way to the Queen Charlotte Islands. High-rise towers in Portland, Seattle, Vancouver, and Victoria begin to undulate. Cascadia's shockwave hammers through sandy soil, soft rock, and landfill like the deepest notes on a big string bass. The mushy ground sings harmony and tall buildings hum like so many tuning forks. The earth rings like a bell as three plates of crust find a new equilibrium.

On I-5, the main north–south interstate highway, thirty-seven bridges between Sacramento and Bellingham, Washington, collapse or are knocked off their pins. Five more go down between the Canada–U.S. border and downtown Vancouver. The most vital overland lifeline from California to British Columbia is severed and bleeding badly. The Trans-Canada Highway has been cut in three places east of Vancouver. All the big bridges spanning the Fraser River in metropolitan Vancouver, around Puget Sound in Seattle, and across the Columbia in Portland have been damaged. None has collapsed outright, but they are considered unsafe until inspection teams can check them out. All major highways leading out of the big cities are plugged with debris from toppled buildings, rockslides, and traffic jams.

Nineteen railway bridges along the north–south coastal mainline of the Burlington Northern Santa Fe railway are wrecked as well. Boulders block the east–west mainline tracks of both the Canadian Pacific and Canadian National railway systems. Three engines and twenty-nine chemical tank cars, at least half of them full of chlorine, derail and spill their deadly cargoes just outside Tacoma. Deep-sea shipping docks in Portland, Seattle, and Vancouver slump and crack, their pilings undermined by liquefaction. As miles and miles of dikes around the city of Richmond, British Columbia, turn to mush, the incoming tide sweeps inland, swamping much of the city.

The runways of every major coastal airport from northern California to Vancouver are buckled, cracked, and no longer flyable. An Airbus on short-final at Sea-Tac touches down just as the concrete breaks. The impact shears off the forward landing gear, causing the jet to belly skid for 120 yards (110 m) before bursting into flames. Dozens of other inbound flights now must find someplace else to land.

Sixteen emergency care hospitals in Vancouver, Victoria, Seattle, and Portland—many of them built before the latest earthquake codes came into effect—suffer full or partial failure of load-bearing walls. The oldest wing of St. Paul's Hospital in Vancouver collapses in a shower of red bricks and dust. The wings that remain standing are only partly functional because the emergency power generators either don't kick in or are running at less than full capacity. More than 580 public school buildings that would have been used as triage and refugee evacuation centers have been badly damaged and are unsafe to enter. There was never enough money to reinforce them all in time.

After fifty cycles of harmonic vibration, dozens of tall buildings have shed most of their glass. In some downtown intersections the cascade of broken shards has piled up three feet deep. Whirling sheets and splinters of broken windowpanes sail down windy canyons, slicing and maiming and killing as they go. The tops of high-rise towers bang together like bull goats in rut.

Shockwaves have been pummeling the Pacific Northwest for four minutes and thirty-five seconds now and it still isn't over. After sixty-four cycles—skyscrapers swaying rhythmically from side to side in giddy wobbles—enough welds have cracked, enough concrete has spalled, enough shear walls have come unstuck that some towers begin to pancake. The same death spiral everyone saw in New York on 9/11 happens all over again. Smaller buildings, but more of them. Dozens go down in the four northern cities. Even as far south as Sacramento, damage to tall buildings is moderate to severe. Glass is falling and people are screaming down stairwells.

In all five major cities tens of thousands of people have been seriously injured. Hundreds, perhaps thousands, more are dead and all the coroners are quickly overwhelmed. But it's too soon to count bodies. More than a third of the oncoming shift of police, firefighters, paramedics, nurses, and doctors do not show up for work. They are either stranded by collapsed buildings, bridges, and roadways, injured or dead themselves, or sticking close to home to make sure their own families are okay before going to work. People who survive the collapses must do their own search and rescue for family members, friends, and neighbors still trapped in the rubble. Help will come eventually, but who knows when?

Tens of thousands of people gather in streets, schoolyards, and city parks, searching for safe ground, in dire need of relocation. Fright, confusion, and panic ripples through the huddled masses. Pets are running loose, barking mad, and there are reports of wild animals escaping from zoos. Looting has already begun and local authorities are quickly overwhelmed. The governors of Washington, Oregon, and California declare states of emergency and call the White House for federal backup. The National Guard is mobilized in all three states.

Canada has no national guard. A handful of coast guard and navy vessels from the Esquimalt base on Vancouver Island are getting mobilized, but most of the active army units are stationed back east or deployed overseas on peacekeeping or combat missions. The engineering battalion and all its equipment has been moved east of the Rockies in a budget-cutting exercise, so it will take many hours or perhaps days for heavy rescue teams to get past the landslides, wrecked bridges, and buckled runways.

When Canada's prime minister calls his good friend, the American president, the news is not encouraging. The United States is committed to several foreign war zones, so there are no heavy-lift transport planes or helicopters readily available to help in British Columbia. Every troop and every spare piece of equipment in operational condition has already been dispatched to Washington, Oregon, and California. All twenty-eight of America's urban rescue teams specially trained to save the lives of people buried under tons of twisted steel and concrete are fully engaged and too busy to help the Canadians. There are five such specialty teams in Canada, but with so many transport lifelines severed, it's going to be damned difficult to get them to the disaster zone in time.

The mountain of water lifted by more than eight hundred miles (1,300 km) of continental shelf suddenly heaving up has now collapsed under the force of gravity into a series of nine tsunami waves traveling east from the subduction zone toward North American shores and west across the Pacific at the speed of a jetliner. As predicted, the first swell of angry seawater hits the beach at Samoa, California, only eight minutes after the earthquake began. Those who survived the devastating temblor but forgot the tsunami drill—or those who simply didn't hear the siren or couldn't move fast enough—are killed almost instantly, battered and drowned as their wood-frame homes now disintegrate and are swept off their foundations.

Moments later the tsunami roars into Humboldt Bay, smashing the waterfronts of Eureka and Arcata. PG&E's old nuclear reactor has been replaced by a conventional, fossil-fueled power generator. When cold seawater hits the boilers, they explode. Two minutes later Cascadia's wave hits the boat harbor at Crescent City, demolishing everything that wasn't already trashed by the earthquake.

Fifteen to twenty minutes later, the same scenario plays out again and again as wave after wave comes pounding across the sand, blasting like some massive, demonic fire hose through the streets of coastal towns like Newport, Cannon Beach, Seaside, and Astoria in Oregon; Ilwaco, Long Beach, and Grays Harbor in Washington; Ucluelet, Port Alberni, Tofino, and Victoria on Vancouver Island. And dozens of other towns and villages in between.

Nearly all the twisty, two-lane highways that connect the coastal communities to the outside world have been buried in several places by mountain rockslides and huge trees. Local fishing harbors, marinas, and docks have all been severely damaged. Only the handful of boats that happened to be at sea this rainy spring morning escape undamaged. Many others get crushed against their docks and capsized, or they drag anchor and grind against the rocks. People who have chosen to live on the edge of this ocean paradise are well and truly on their own now. It may be a week or more before outside help can get here.

The westbound tsunami waves, meanwhile, continue to hurtle across the Pacific, fanning out in wide arcs that will make landfall with sledgehammer force in hundreds more coastal villages, towns, and cities. Hour by hour as the waves get closer, alarms sound in more than a dozen languages and dialects. Hawaii, Midway Island, Alaska, and the Kamchatka Peninsula are among the first to be struck. Although physical damage is heavy, the loss of life and injuries are kept to a minimum because NOAA's computer model has accurately predicted exactly how big the waves will be, which beaches will be hardest hit, and when the tsunami will arrive. The evacuations are largely successful.

Roughly nine hours after the quake, Cascadia's first wave hits Japan's eastern seaboard. Here most people have moved to higher ground in time. But the damage to waterfront homes and villages, and especially to high-tech container shipping docks, is extensive. Some of the busiest high-volume, high-value commercial shipping terminals in the world are dealt a severe blow and knocked out of business for who knows how long. Much of Japan's export trade is essentially crippled.

The same thing happens again and again as Cascadia's waves crash ashore in the Solomon Islands, New Guinea, the Philippines, Taiwan, and Hong Kong. After that it's onward to Indonesia, Australia, New Zealand, and the western shores of South America. The waves, like those generated off Sumatra in 2004, actually turn corners at the bottom of the planet. Eventually the tsunami dies, exhausted, in the Indian and Atlantic Oceans and on the frigid shores of Antarctica. Before the day is done seven of the world's largest insurance companies file for bankruptcy. There's absolutely no way they can pay all the claims.

Back in 2005 a group of scientists, engineers, and emergency planners from the United States and Canada, along with representatives from key industries in the Pacific Northwest, formed a committee called CREW—the Cascadia Region Earthquake Workgroup—to study the magnitude and severity of problems posed by the coming event. Their official disaster scenario states that the rupture of Cascadia's fault "could be catastrophic . . . It will be a long-term event, affecting the economies of the US, Canada, other Pacific countries, and their trading partners for years to come."

In my own, hypothetical version of this day of reckoning, they were absolutely right. Although nobody has ventured an official guess at how high the death toll will be worldwide, most experts agree that Cascadia won't kill as many as Sumatra did simply because not as many people live right on the beach. Generally speaking the homes and cities of the Pacific Rim are built a bit more solidly than their counterparts around the Indian Ocean. But the cost of repairing or replacing the damaged or destroyed infrastructure of the Pacific Rim will probably be many times higher. Modern cities cost more to build in the first place and much more to fix when they get smashed. So the economic consequences of a Cascadia quake will be like nothing we've ever seen. Some say it will take a decade or more to dig ourselves out of the rubble. From a purely dollars-and-cents perspective, the whole world will feel our pain.

But CREW's conclusion is not entirely bleak or defeatist: "A Cascadia earthquake will seriously affect our region, but it won't destroy us. We will rebuild our cities, our neighborhoods, and our businesses. The time it takes us to recover will depend largely on what precautions we take before the earthquake." Here again, I think they've got it absolutely right.

EPILOGUE
Survival and Resilience, a State of Mind

Since I began work on this book, tectonic events have made scary headlines five more times. A team of seismologists in Italy has been threatened with charges of manslaughter for failing to predict an earthquake that they allegedly saw coming. After several tremors were detected in March 2009 the nation's Major Risks Committee, a scientific advisory group like the "six wise men" in Japan, met to discuss whether anything in the data could be classified as a reliable precursor to a quake.

After the meeting a government official told reporters the scientists had concluded there was "no danger, because there is an ongoing discharge of energy" being released by the small tremors. Then on April 6, 2009, an earthquake of magnitude 6.3 struck the city of L'Aquila, killing more than three hundred people and injuring sixteen hundred others. The threat of manslaughter charges against members of the advisory committee caused an uproar among earthquake researchers around the world.

As of September 2010 almost four thousand scientists and engineers had signed a letter to the president of Italy calling for an end to what some termed a witch hunt. They urged the government to spend more resources on "earthquake preparedness and risk mitigation rather than on prosecuting scientists for failing to do something they cannot do yet—predict earthquakes." Barry Parsons, an earth scientist at the University of Oxford and one of those who signed the letter, explained that "scientists are often asked the wrong question, which is 'When will the next earthquake hit?' The right question is 'How do we make sure it won't kill so many people when it hits?'"

In Haiti on January 12, 2010, a magnitude 7.0 earthquake destroyed huge sections of the capital city of Port-au-Prince, killing more than 200,000 people. The island of Hispaniola sits on the rim of the Caribbean plate, which is being shoved westward by the North America plate as it dives underneath. But here again the real story was about the exposure to risk caused by poor construction quality. Buildings collapsed and people died not for lack of warning but because poverty and shabby construction practices made the tragedy inevitable.

Little more than a month later, on February 27, 2010, another big subduction earthquake struck the coast of Chile. At magnitude 8.8 it released approximately five hundred times more energy than the Caribbean shock yet it killed far fewer—roughly seven hundred lives were lost in Chile compared with the 200,000 in Haiti. Observers commented that construction quality in Chile was definitely better because building codes have been strictly enforced ever since their last big quake in 1960—at magnitude 9.5 still the largest rupture ever recorded.

The February jolt occurred on the same fault, twenty-two miles (35 km) beneath the sea floor, and apparently picked up right where the last one stopped. It ripped the next four-hundred-mile (640 km) segment of the subduction zone and probably relieved most of the remaining stress built up in the system. That was the good news. But there was more.

Because the 1960 quake did not rip the entire fault, scientists expected the next segment to go at any time. So several groups of researchers had installed an extensive array of GPS monitors to keep track of the strain build-up. Computer models (including one by Kelin Wang at PGC on Vancouver Island) then made predictions of how much the fault would move when the last segment finally did break.

On February 27 the predictions turned out to be right on the money. Almost ten feet (3 m) of nearly instantaneous horizontal motion was measured by GPS instruments at Concepción. The land lurched sideways nearly nine feet, just like Wang's model projected. Which is the same kind of movement Herb Dragert and Mike Schmidt expect to see in Victoria when Cascadia rips loose. Scientists can now be fairly confident in predicting which part of a fault is likely to break, how far it will move, and how large the jolt will be.

But in Chile other positive factors were at work as well. One of them was a higher level of public awareness. Because enough people remembered 1960, they knew what to do when the ground started to rumble. Those living in danger zones near the beach immediately ran to higher ground without waiting for someone in authority to issue an evacuation order. According to Kelin Wang, "Numerous lives were saved by this kind of self-evacuation . . . The importance of educating the public far exceeds that of warning buoys."

And then it happened again. On Tuesday, February 22, 2011, a strike-slip fault near Christchurch, New Zealand, ruptured in a magnitude 6.3 earthquake that killed more than a hundred people outright and buried hundreds of others in rubble. This time two tectonic plates ripped past each other horizontally, like the San Andreas plates had done. So it wasn't a subduction quake like the ones in Chile or Alaska or Cascadia (no heaving up of the ocean floor and no tsunami), but it was a rupture in another corner of the Pacific region's infamous Ring of Fire.

Less than a month later, on Friday afternoon, March 11, 2011, another segment of the Ring of Fire tore apart in a magnitude 9 earthquake that did jack up and shift the ocean floor off the northeast coast of Japan near the city of Sendai. The shockwaves lasted between three and five minutes and caused skyscrapers in Tokyo—231 miles (373 km) away—to sway like trees in a strong wind. The tsunami was seen around the world almost instantly, covered live by Japanese television crews in breathtaking and heartbreaking detail for hours on end. We were all able to witness immediately the apocalyptic extremes of seismic chaos.

Rolling balls of flame and thick smoke billowed from ruptured tanks at an oil refinery. A farm family's home burned furiously as it was carried away—floating atop a tangled mat of splintered lumber and logs, twisted sheets of metal from busted barns and sheds—all of it swept across the coastal lowlands on a thirty-foot (10 m) wave, a churning tsunami gumbo of black soil and seawater. Large commercial fishing boats were torn from their moorings and tossed against concrete breakwaters. The boats bounced off the breakwaters and then slammed like battering rams into the walls of nearby buildings. Hundreds of cars and trucks were carried away on this roaring tide of muck, right before our eyes. Were there people inside? How could there not have been?

Mud-spattered survivors looking stunned and forlorn wandered through rubble in search of family and friends. Hundreds of people sprawled, exhausted and in shock, on the floor of a school gymnasium, nesting in rucked-up blankets, their coats and shoes and "grab-and-go" bags gathered around them like imaginary walls to fend off the ongoing nightmare.

Food store shelves were stripped bare and long lines of cars appeared at gasoline pumps, the fuel supply running dangerously low. And still, somehow, all of this post-quake scramble was happening in a relatively organized and orderly manner, presumably because the Japanese had experienced many smaller quakes and tsunamis before and knew this was coming. They had planned and drilled and rehearsed. The atmosphere seemed amazingly calm and eerily quiet. I sincerely doubt it will be this peaceful when the same thing happens to North America.

But if pride comes before a fall, then the construction of nuclear reactors near an active fault zone has to rank among the most dazzlingly optimistic—or stunningly foolish—things that modern nations have ever done. Not just in Japan, but in California and many other places around the globe. Until we watched those plumes of gray-white smoke rising across the Japanese coast as the roof and outer walls of the Fukushima Daiichi reactor complex began to vaporize and then collapse—until we saw it with our own eyes—the horror of a nuclear meltdown seemed like the last thing a quake or tsunami survivor should have to worry about. Scientists, engineers, and government officials have led us to believe that nuclear plants are built to withstand seismic shocks. Now I guess we know better. For those living on North America's own locked and loaded segment of the Ring of Fire, the question now must be: are we next?

Even though seismologists still cannot predict when Cascadia's fault will break, pretty much everybody who has lived out west for a while knows deep down inside that a megathrust quake will eventually happen. Sure, on any given day the mathematical odds of a magnitude 6 or 7 under downtown Seattle or Vancouver are higher than for a magnitude 9 from Cascadia. Smaller quakes do happen more often than big ones. That's why most emergency managers have been told the local rumble is still their worst scenario. And for any given point on the map, that's absolutely right—the local quake may cause more intense damage to that particular city than Cascadia would.

But Cascadia's fault is going to cause damage to *all* the cities and towns along a swath more than 800 miles (1,300 km) from north to south and as much as 125 miles (200 km) inland. So the cumulative damage will be far greater than the impact of any local quake on any single city. The enormity of what's about to happen in the Pacific

Northwest is almost inconceivable. And that's only part of the reason why Cascadia is not yet as infamous or worrisome to many people as the San Andreas already is.

The main reason why emergency managers and even elected officials tend to focus almost exclusively on their own local concerns is that their jurisdictions demand it. As Lori Dengler at Humboldt State University pointed out, the Governor's Office of Emergency Services in California does not have a mandate to worry about what might happen in Oregon or Washington, much less the consequences for British Columbia. But as soon as Cascadia breaks we're all going to be out there in the rubble together, and that's a hell of a time to get to know each other.

If the biggest, wealthiest, most technologically advanced nation in the history of the world could not cope with Hurricane Katrina any better than it did, how on earth will it cope with Cascadia? How will Canada? And here's the thing: with Katrina there was at least forty-eight hours' worth of muscular wind and howling rain before the main part of the storm hit New Orleans. When Cascadia's fault ruptures there will probably be no warning at all.

So what should we do—slit our wrists? Absolutely not. Even after more than twenty-five years of watching people ignore the obvious, I still think we can survive this thing better than we might imagine. As Patrick Corcoran said time and time again—yes, it is likely to happen; yes, it will be bad; and yes, education can make it better. So go out there and get some. Join an emergency preparedness group. Take a first aid course. Just get up off the couch and do something.

Bottom line: we're *not* all going to die! The vast majority of us will survive the big jolt from Cascadia. The key issue is how well we endure the aftermath. And that depends totally on how much time and attention we invest now in preparing ourselves, making our communities resilient.

Two anecdotes told by Lori Dengler when we interviewed her for the 2008 *ShockWave* documentary stick in my mind and give me something to hang onto. A bit of hope. After the Sumatra earthquake and tsunami Dengler traveled to the disaster zone to study what had happened and how it might apply to us in North America. Like Chris Goldfinger and others she came back to her laboratory on the north coast of California with a huge volume of new data and insights about how a subduction quake works and what a tsunami will do.

But the story of Tilly Smith was one she simply had to tell anyone who would listen. The film crew and I definitely listened. Tilly Smith, a ten-year-old British schoolgirl on Christmas holidays with her parents in Thailand, was strolling across the sands of Mai Khao Beach near Phuket when she noticed frothing bubbles on the surface of the sea as the tide started to recede quite suddenly. Two weeks before Christmas break Tilly had learned about tsunamis in her geography class. Old film footage of a wave that hit Hilo, Hawaii, back in 1946 had evidently left an indelible memory because she immediately recognized the same thing and ran to tell her parents.

"I told my mom again and again," she squealed later in a television interview, "and I was hysterical at this moment, saying, you know, 'There's *going* to be a tsunami! There's *definitely* going to be a tsunami!' You know? Just *believe* me!"

"Her mum and dad *did* believe her," said Dengler, "and they managed to clear everybody off the beach. Got them into the hotel. And they managed to vertically evacuate and get about a hundred people from that hotel into the upper floors. And not a single person died in that particular hotel complex. All because a ten-year-old girl had knowledge! All because she recognized the natural warning signs."

Even more to the point of our survival in Cascadia's shadow was Dengler's story from a personal visit to Simeulue Island, a tiny tropical outpost roughly ninety miles (150 km) off the west coast of Sumatra: one of the nearest human settlements to the epicenter of the magnitude 9.3 earthquake of Christmas 2004.

"Simeulue Islanders are a relatively homogeneous people," Dengler explained. "They're still very much in touch with their tribal identity and a strong oral tradition." Legend has it that a quake and tsunami struck Simeulue in 1907, killing many local residents. Those who survived evidently told this story to their children and grandchildren, which may have been why they knew what to do when the same thing happened again in 2004.

"Langi village was the village closest to the epicenter of that earthquake," Dengler continued. "They felt that earthquake very, very strongly. And in fact it damaged about 25 percent of their structures. The first tsunami waves arrived at that northern part of Simeulue Island only eight minutes after the earthquake," she said. "They had very, very little time."

But because their oral history had been kept alive, they knew exactly what had to be done. According to Dengler, everyone in Langi knew that "when you feel a really long strong duration earthquake, you immediately grab your children, help grandma—and get yourself up to high ground. And not only did they go up to high ground, they actually had an entire temporary village—materials to *make* an entire temporary village up there. They had posts; they had aluminum for roofs; they had water; they had food." She sounded impressed, almost amazed. And so was I. Best of all, however, was the conclusion to her story. After the earth shock came the waves.

"The waves were enormous," said Dengler. "In Langi village every single house was completely wiped off the face of the earth. The only thing left were the concrete foundations." She shook her head. "Completely destroyed. They lost their animals; they lost their fields." But? And I knew there had to be a but. "Not a single man, woman, child— not a single old person—died," Dengler said. "Not one!"

Thinking about it now I'm pretty sure I had to look away. We both had tears in our eyes. "Oh, it was an amazing story," she continued. "To me the most important lesson from Indonesia is that if you have an aware community, you can all survive. But you need to keep that as a part of your culture. You need to make sure that it's not forgotten from generation to generation."

And now we on the west coast of North America must learn to do the same. Dengler put it in perspective: "Most people here are going to survive a Cascadia event. But a Cascadia event is going to have more people having to be on their own and self-reliant and resilient for a longer period of time than any other event that I can think of." So there's more to it than duck-and-cover drills in schools and tsunami evacuation exercises.

We have to gather at the neighborhood level and in family groups to come up with personal survival plans that instantly come to mind no matter where we are when it finally happens. With busy lives that seem like a game of musical chairs, we need to know what we'll do wherever we happen to be when the music stops: at home, at work, or at play, as Patrick Corcoran likes to say.

We also need to think hard about mitigation, about renovating and upgrading essential structures such as hospitals, schools, and other public buildings that we'll need for emergency shelters. Computer models have already shown some communities that their police stations, fire halls, and other important infrastructure may be wiped out in the coming flood. City and village councils should already be thinking about how and when (and how to afford) to move some of these facilities or rebuild them on safer ground. Governments need to pass legislation to rezone the dangerous parts of town so that if public or private buildings are destroyed by Cascadia's fault, people will know ahead of time that they cannot and should not expect to simply rebuild in the same spot, re-creating the same fatal vulnerabilities.

Eddie Bernard of NOAA summed it up nicely. "Don't we as a society want to save the *community*?" he asked. "That is—you want to have a community to return to. You want to have a hospital to go to. You want to have schools that your children can go to. You want to have teachers in those schools. All those things were wiped out in Sumatra. They lost everything. So that's a lesson that we should take away—we should be building our society so that it's resilient to the next tsunami." And even though the job sounds daunting, Bernard remains an optimist. "What we don't want to have is just the assumption that it's hopeless. Because it's not."

I agreed. And so did Chris Goldfinger: "It doesn't have to be such a disaster. It's only a disaster if we don't do something."

AFTERWORD

Lessons from Tohoku—A "Worse Than A Worst-Case Event"

At magnitude 9.0 the Tohoku-Oki earthquake on March 11, 2011, was the strongest ever to hit Japan. The main shock was followed by more than nine hundred aftershocks, sixty of which measured magnitude 6.0 or higher. Three of them reached magnitude 7.0 or higher. An estimated twenty thousand people are dead or missing, more than 95 percent of whom were killed not by the quake but by the tsunami that followed. The U.S. Geological Survey (USGS) estimates that 332,395 buildings were either destroyed or damaged. At a cost of more than $309 billion, this seismic event has become the most expensive natural disaster in history. And it all began with a quake exactly like the one Cascadia's fault will generate along North America's west coast.

At home in British Columbia, it was Thursday night, March 10, 2011. In Japan it was Friday afternoon. Out of the blue I received an urgent email from Chris Goldfinger of Oregon State University.

It was a madly dashed mass-mail-out to his wife, colleagues, and friends around the world. It basically said that he was okay, that he had survived the quake, and that he would tell us more as soon as he could.

At that very moment, I was watching coverage of the largest tsunami in a thousand years as it rolled over the top of a thirty-three-foot-high (10 m) seawall in northern Japan.

Goldfinger just happened to be in Tokyo attending a science conference about the Sumatra earthquakes of 2004 and 2010. "It was the third earthquake of the week," he blogged a short time later. Japan had been rattled by a magnitude 7.4 earlier in the week and a smaller one the next morning, both of which may have been foreshocks.

The jolt on Friday afternoon, however, was definitely the main event. "At first, it seemed no larger than the others . . . But instead of stopping at a few seconds, or maybe a minute as the earlier ones had, this one kept going, and going, and going.

"In a room full of seismologists, we timed the gap between the P-wave and S-wave arrivals, and then started thinking about whether we should get out of the building. The desks looked really flimsy, so duck and cover didn't look good at all . . . After about a minute of shaking, we were all outside in the courtyard watching the flagpole on the roof of the seventh floor whipping through sixty degrees. And the dry rattle of the trees with last year's leaves as they shook . . .

"The main shock lasted an eternity . . . I never realized you could feel the difference between the different types of waves. The P-wave is like a jackhammer under your feet; the S-wave much more like an ocean wave. We all felt a little seasick . . . The aftershocks were nearly continuous for the next twelve hours or more. It's a long time for the earth to feel like the ocean."

In a television interview months later, Goldfinger explained that the quake was far bigger than anything anticipated from the Japan Trench. There was befuddlement in the science community about how conventional wisdom could have been so wrong. He wanted to know "how this earthquake could have happened in a place where, according to our pet geophysical theories, it should have been impossible."

Nearly a year later, the first flurry of scientific papers began to emerge, analyzing what went right or wrong and what lessons had been learned from the Tohoku quake. I read Goldfinger's first messages again and tried to put myself on the ground over there—to think of what it must have been like as the earth came apart at one of its deep-ocean seams.

I pictured myself on a street in a fishing town called Minamisanriku, on Japan's northeast coast, to imagine what I might have done when the ground began to shake at 2:46 p.m. I'm pretty sure I would have dropped to my knees as the shockwaves started rolling through the sidewalks. As buildings started to rattle and sway, I would have heard glass breaking. Sirens and loudspeakers blaring. People spilling into the streets—not in a panic yet, because they've been through this drill so many times before. The first big fist of seawater would not arrive for twenty or thirty minutes, so there would still be time to respond.

The big question would be what to do next. Would I perceive myself to be at risk and head for high ground in the dark, forested hills beyond the main village? The safe zone would be in plain sight. It would be easy to run there in twenty minutes. Or would I feel fairly secure hiding behind the massive concrete seawalls that stood between humanity and the roiling-mad North Pacific?

If I were a local citizen, I would also know exactly where to find at least one of the eighty buildings in this village designated as tsunami-evacuation shelters, so I would probably make a beeline for the nearest one instead of jogging all the way to the hills. And why not? Government officials had assured everyone that these buildings would be safe places to escape a killer wave. What could go wrong?

Minamisanriku had earned an international reputation for being pre-pared for the worst disaster imaginable. A small army of scientists had studied the thirty-two biggest earthquakes (from magnitude 7.0 to 8.5) that had rumbled through the region since 1900 from a subduction zone called the Japan Trench, 124 miles (200 km) offshore. Others had com-puter-modeled the highest tsunami wave conceivable based on the "most credible earthquake." They were pretty sure they knew what was coming.

For decades the government of Japan had braced itself by spending billions of dollars to build dikes and breakwaters along 30 percent of the eastern seaboard. These concrete barriers were up to sixty-six feet (20 m) thick, anchored to the ocean floor nearly 200 feet (60 m) down, and in some cases rose as high as thirty-three feet (10 m) above the sea surface—surely high enough to deflect any doomsday wave that might come along. It would be hard to find a more organized, drilled, and solidly grounded community in the most earthquake- and tsunami-ready nation on earth. If anybody knew how to cope, it would be the people of Minamisanriku.

Now in my mind-movie, I can see several small boats scurrying to get inside the harbor before the giant tsunami gates grind shut against the outer bay. A team of firefighters stands by as the steel doors clang together to form a solid barrier against the coming waves. I've seen footage of this scenario on television several times. The people were proud of how quickly they could seal themselves off from the ocean's fury: fifteen minutes max. And on March 11, 2011, that ponderous machinery worked exactly as planned.

Unfortunately, the plan was the problem. It was no match for what nature dished out. The quake was much, much stronger than expected, and the waves much higher than any tsunami in living memory. The subduction zone did something scientists had seen in computer simulations but never before in real life.

For the first sixty seconds, the fault ripped apart pretty much as expected. The overlying tectonic plate—the Okhotsk plate, upon which Japan's main island of Honshu rests—came unstuck from the Pacific plate, the slab of rock that was pushing its way underneath. From the epicentral area, some nineteen miles (30 km) below the sea floor, the lower plate at first lurched downward and westward—toward Japan. And then it kept on ripping. Segment after adjacent segment of the Japan Trench broke apart, spreading damage farther and farther along the coast. The earthquake kept getting bigger. What started at magnitude 7.9 quickly jumped to 8.8.

After seventy-five seconds, the fault did a stunning turnabout: it started rupturing in the opposite direction. In two directions at once, actually. Now it tore itself apart from the epicenter *uphill* along the fault plane and eastward, farther offshore—out toward the Japan Trench where the two plates first meet.

The world's most sophisticated and densely packed grid of seismometers, pressure gauges, and GPS tracking devices measured the steadily increasing magnitude of the quake and the height of the coming waves and beamed the data to satellites. The data automatically triggered alarm systems in Japan and around the Pacific Rim. Bullet trains screeched to a halt, and nuclear power plants initiated emergency-shutdown procedures.

It took more than 180 seconds for the rocks to release several centuries' worth of stress. An area of sea floor roughly 250 miles (400 km) long and 125 miles (200 km) wide lurched sideways—at least 79 feet (24 m) in most areas and as much as 164 feet (50 m) in others. The Tohoku-Oki earthquake will probably turn out to be the largest fault slip ever observed. It was more than twice the peak slip recorded during the Sumatra quake of 2004 (magnitude 9.4). It had twice as much horizontal movement as seen in the Chile quake of 2010 (magnitude 9.0). It caused an even bigger lateral lurch than the one that occurred during the Chile disaster of 1960, the largest earthquake ever recorded (magnitude 9.5). How could this be?

Out along the Japan Trench, the thin edge of the overlying continental plate popped loose from the plate below and heaved upward ten feet (3 m), hoisting the ocean with it. Think of a volume of seawater 250 miles long, 125 miles across (400 by 200 km), and several miles deep being thrust upward—all that water—roughly ten feet straight up. A seething liquid hump of kinetic energy.

Moments later, gravity forced this saltwater mountain to collapse. A series of concentric waves rippled outward, as if an enormous boulder had been dropped into a shallow pond. The birth of a killer tsunami.

As stress was released from the rocks, the upper plate relaxed. The land stretched like a piece of taffy, then sank lower into the sea than it had been before, dragging the coastline down with it by at least three feet (1 m). From this moment on, many of Japan's outer beaches, bays, towns, and villages would be three feet below sea level.

Thirty minutes after the main shock, thousands of people in Minamisanriku had scrambled into buildings they thought were safe. But because both the quake and wave were much bigger than expected—and because the entire coastline had just sunk by three feet—those "invincible" seawalls simply vanished beneath the surge. The breakwaters and tsunami gates were overtopped and undermined. It was as if they had never been built. All that money and years of effort had created nothing more than a house of cards, a fatally false sense of security.

An estimated 95 percent of Minamisanriku was destroyed. Of the eighty buildings officially designated as tsunami-evacuation shelters, thirty-one were completely wrecked. Two evacuation sites, one on a headland overlooking town from the south and another farther inland, were also inundated; both were at least sixty-six feet (20 m) above sea level and presumed high enough to be considered safe ground. But the people who gathered there were washed away.

The town's mayor and more than 130 public officials escaped from city hall to the Disaster Management Center, a solidly built steel-frame structure three stories tall with a vertical evacuation shelter on the roof. Only thirty made it to the top. Even there they were not safe.

From her desk on the second floor, twenty-five-year-old Miki Ando, an employee of the Crisis Management Department, continued broadcasting emergency information via the city's extensive network of loudspeakers; she never left her post even as the horrifying waves tore through the streets below her and climbed ever higher against the building's groaning walls. Ando repeatedly urged her fellow citizens to head for higher ground, and she was credited with saving many lives by doing so.

A time-stamped photo taken from the top of the building and discovered days later showed that the Disaster Management Center was fully engulfed by 3:35 p.m., forty-eight minutes after the earthquake had started. Of the thirty people who made it to the roof, only eleven survived, several of them clinging for life to a radio antenna as the ocean torrent—a wall of debris-filled seawater fifty-two feet (16 m) high—thundered past.

An aftermath photo, taken weeks later by geologist Lori Dengler of Humboldt State University, showed nothing left of the building but a gutted steel skeleton. Clearly visible in the background behind this mangled red-metal hulk are the green, tree-clad hills that could have saved more lives if only people had run a few minutes farther and higher.

Miki Ando was among the 9,500 people of Minamisanriku either killed outright or swept out to sea and presumed drowned. More than half the town's population was gone in less than an hour. The same story was repeated in town after town along Japan's battered northeast coast.

When the main shock ended, the final magnitude measured 9.0, the largest earthquake ever to hit Japan. "They were expecting an 8.2 to an 8.4, which is twenty to twenty-five times smaller than a magnitude 9, in terms of energy," explained Chris Goldfinger. Japan had prepared itself for the wrong earthquake.

The waves broke records as well, with a high-water mark of 127.6 feet (38.9 m) at Aneyoshi Bay south of Miyako City. That's a bulldozer blade of water more than twelve stories high. It was the largest tsunami ever measured in Japan's long experience with killer waves. On the broad, flat Sendai plain, the surge was about thirty feet (9 m) high and it penetrated roughly two miles (3 km) inland, destroying farms and homes as far as the eye could see. Japan's northern network of seawalls and breakwaters was designed for a worst-case wave of about thirty-three feet (10 m) above mean sea level. But the ground had sunk three feet, so most of the barriers were "severely overtopped or destroyed," according to a report from the American Society of Civil Engineers (ASCE). They were torn apart by waves that, in many cases, were double the expected height.

Japan's real-time warning system—triggered by a dense array of instruments connected to a supercomputer that can process thousands of potential disaster scenarios—worked, but not as well as expected. Eight seconds after the primary wave from the earthquake reached the first seismic station, the Japan Meteorological Agency (JMA) issued an alarm to people living closest to the epicenter. Almost instantly, twenty-seven bullet trains in the region were stopped without a single derailment.

The computer initially reported a quake of magnitude 7.9 and forecast tsunami surges of about ten feet (3 m) along the Iwate and Fukushima coasts, with waves as high as twenty feet (6 m) in the Miyagi district. This news flashed across the country via television, radio, cell phones, and community loudspeaker systems. But the quake kept getting bigger, and the news kept changing. JMA had to update the bulletin continuously over a period of nearly four hours. In the Greater Tokyo Area far to the south of the quake—where high-rise towers were quite obviously swaying as damaging shockwaves rolled through the ground—there was no official warning. Fortunately, the bullet trains and nuclear power plants in the Tokyo region had their own warning systems to shut the operations down.

Clearly, the supercomputer underestimated the size of the catastrophe. Critics later called this a failure of the system. But the initial notification was based on just the first twenty seconds of data; only part of the fault had ripped within that time, and the computer did not know how much bigger the quake would eventually become. It took nearly three and a half minutes for the whole thing to unzip.

"If you analyze it using twenty seconds of data, you're only going to see about a hundred kilometres [62 miles] of rupture," explained geologist Lori Dengler. "Because that's how long it takes to rupture that amount. By definition, if you're trying to do an early-warning system based on the initial part of the signal, you are always—100 percent of the time—going to grossly underestimate the magnitude of the event."

In a post-mortem published in *Nature*, Masumi Yamada of Kyoto University wrote that "the unexpected character of the seismic data at the start of the earthquake fooled the early warning system's algorithms." But he added that "the system has the potential to work well for the next great earthquake . . . if technical improvements are made to recognize great earthquakes quickly."

JMA has already revised its warning protocol. From now on, if a quake hits magnitude 8 or higher, the system will not even try to forecast tsunami-wave heights. Instead, bulletins will simply announce the possibility of "a huge tsunami." The question remains whether people will know what that means and what they ought to do. Will they perceive themselves to be at risk and take appropriate action?

Dengler, one of the leaders of an international team of scientists who rushed to the scene in search of clues about what happened, summed it up this way: "Certainly the sirens were going off, and they were announcing it as a worst-case event. But a worst-case event was not what they got. They got a *worse*-than-a-worst-case event."

Perhaps more worrisome to Dengler were reports that nearly 40 percent of people intentionally delayed their own escape to safety. "The earthquake was a trigger [for people] to go from a safe area back into a hazard zone," she told me. "And they did that to either rescue a relative who was at home, or to rescue belongings. And this was common. It was not rare. It was really common.

"And that's the thing that I'm really concerned about. Because I expect we're going to have people doing the same thing *here* [in North America]. The main motivations are to rescue *somebody* or to rescue *stuff*. And that's pretty instinctive human behavior."

Not all the news was bad, however. In some places, people did know what to do and lived to tell the story. More than five hundred students from the junior high and elementary schools in the town of Unosumai managed to escape despite conflicting information and a rapidly changing disaster scene. Both schools lie outside the mapped tsunami hazard zone of their community, so the students and teachers there might be forgiven for thinking they were relatively safe to begin with.

The two groups of students had been told completely different things about what to do in an earthquake. The elementary-grade children were trained to go to the roof of their three-story school building. The neighboring junior-high students had been taught to head for higher ground. But when the earth ripped apart on March 11, the younger students saw the older ones leaving the school grounds and decided to follow.

The older kids noticed, took charge of the younger ones, and led the way, always keeping an eye on the churning pulse of dirty-brown seawater rising behind them at a terrifying speed. They had to change their planned destination twice before finally reaching a safe spot in the nearby hills. Coaching the students to head for high ground "to evaluate the situation with their own eyes" and to assist others clearly paid off. All five hundred survived.

Most people who did make it to official evacuation sites were stranded for days. Few of the shelters had stored supplies of food, water, blankets, or bedding. There were not enough toilets and, in many cases, no first-aid stations or emergency medical care. Mid-March in northern Japan means winter weather with temperatures hovering around the freezing mark. In some cases, elderly and injured survivors later died because of the difficult conditions at the evacuation sites.

"There was no catastrophic-response plan," explained Lori Dengler. "There was no idea that you'd really need to coordinate response on a nationwide level. There was no plan because everything was built on the faith that the engineering works were going to work."

An estimated 92 percent of the nearly twenty thousand people who did not survive died as a result of drowning. If you count those who were washed out to sea and never found, then 96 percent of the deaths were caused by the tsunami. A significant number of these unfortunate people died because they had put their faith in seawalls and official disaster plans. The best of intentions had undermined decades of diligent training, preparation, and public awareness.

In November 2011 the Earthquake Engineering Research Institute (EERI), based in Oakland, California, published a summary of the findings from several field reconnaissance teams sent along with two International Tsunami Survey Teams (ITST) to the hardest-hit areas of Japan. In essence the EERI teams concluded that "failure to evacuate" was the primary cause of the high casualty rate. They also noted that the official forecasts had underestimated the strength of earthquakes along the Japan Trench because the projections were based on what had happened during "relatively recent" historic events.

The scientists who made those projections quickly admitted that "the historic record" was indeed where the trouble started. Two months after the Tohoku disaster, five prominent Japanese seismologists wrote a post-mortem (published in *Nature*) to explain what they had learned from the tons of new data gathered in the aftermath. One of the authors, James Mori of Kyoto University, was remarkably candid in his assessment: "The Tohoku earthquake came as a frightening and disheartening surprise to Japanese seismologists."

Most experts had thought it should be possible to predict the locations and approximate magnitudes of these monstrous plate-boundary earthquakes. They gathered the existing geologic data and did the math. But it turned out that roughly four hundred years of historic records covered "too short a time period to be a reliable guide." Nothing larger than a magnitude 7.5 had been seen since 1923. There was no written record of anything larger than an 8.5 since the seventeenth century. A long-term forecast for the Tohoku region issued by the Japanese government in 2002 estimated an 80 to 90 percent probability that the area would have an earthquake of magnitude 7.7 to 8.2 sometime in the next thirty years. The probability "of a magnitude 9 earthquake affecting a 400–500 kilometre [250–310 mile] area was not specifically mentioned," wrote Takeshi Sagiya of Nagoya University. "As a member of the working group involved in the evaluation, it is with great regret that I reflect on the causes of this failure."

For several years there had been discussion and debate about how much pressure was building up along the subduction zone. Japan's land-based network of GPS tracking devices had revealed that the tectonic plates were converging at an overall rate of nearly four inches (9 cm) per year. A rather frightening speed when it comes to potential earthquake generation, especially when you think of four inches multiplied by hundreds of years. It seemed that there was more stress going *into* the rocks along the fault than there was coming *out* during earthquakes. The numbers didn't add up. The term "slip deficit" came into use. Slippage along the fault during all known earthquakes in historic time had not been enough to bleed away the huge amount of pressure that should be there if two giant slabs of the earth's crust were converging this quickly.

As far back as 1996, Yasutaka Ikeda of the University of Tokyo had warned that "the strain accumulated in the last 100 years at abnormally high rates is likely to be released by slip on the megathrust at the Japan Trench, which would produce big earthquakes with magnitude 8 or greater." And *greater* is exactly what happened on March 11, 2011.

The energy released during the Tohoku quake was astonishing. "The amount of strain release is nearly an order of magnitude larger than what we have seen in other megathrust earthquakes," wrote Hiroo Kanamori of Caltech in *Nature*. "The strain must have accumulated in this zone for nearly 1,000 years." And clearly not all of it had been released by smaller quakes during that time.

Apparently the same thousand-year timeline was true of giant tsunamis. Waves as large or larger than Tohoku's had hit Japan's east coast before. Scientists from Tohoku University in Sendai, led by geologist Koji Minoura, dug trenches and took core samples of sand deposits from a tsunami triggered by a shockwave, now known as the Jogan earthquake, that occurred on July 13, 869. The waves penetrated more than two miles (4 km) inland, so the Jogan quake produced a tsunami as large or larger than the one we saw in 2011.

But that's not all. The researchers found at least two more sand sheets buried underneath the Jogan layer, each of them about the same thickness as Jogan's and each representing another huge wave. Which means that the same astonishing release of seismic energy has happened three times in roughly three thousand years, according to the findings of Koji Minoura and his colleagues. Their paper, published back in 2001, concluded that large-scale tsunamis happen every 800 to 1,100 years. They pointed out that more than 1,100 years had passed since the Jogan event and that therefore "the possibility of a large tsunami striking the Sendai plain is high." Thus, they sounded the alarm ten years before the Tohoku disaster of 2011. But like Ikeda's caution in 1996 about the likelihood of high stress values on the fault triggering large earthquakes, their words of warning somehow failed to reach the right people.

Calculations made by various agencies to forecast the next big quake and tsunami on Japan's east coast did not include Jogan or the other, more distant events of the prehistoric past. Perhaps with so many smaller jolts in recent centuries, those who did the math simply assumed they had enough data to figure out what was "most likely" to occur on a purely statistical basis. Smaller quakes do happen more often, and therefore they are thought more likely to occur.

Three main events in *recent* history—in 1896, 1933, and 1960—had convinced officials that their seawalls and tsunami gates would be strong enough and high enough to defend the coast. They had enough data—geologic and historical—to form an image of the maximum credible event (MCE), the earthquake and tsunami most likely to occur along this part of the coast. Something like a magnitude 8 must have seemed like the logical thing to plan for.

More than two dozen ancient tsunami marker stones had been found in northeastern Japan. One marker stone in the hills above Aneyoshi

Bay, showing how high the water from the Sanriku tsunami of 1896 had reached, had an inscription suggesting the village be rebuilt above that level. Considering that 22,000 people died in the 1896 tsunami, the survivors did decide to relocate their community above the high-water mark. And fortunately, in 2011 the highest Tohoku wave fell thirty-three feet (10 m) short of the 1896 marker. The people who had resettled above the stone were indeed safe. However, a survey by the ITST investigators showed that at least 20 percent of the other marker stones in the area had been overtopped or washed away.

While the Tohoku tsunami was the highest in more than a thousand years, the Jogan wave pushed seawater at least 650 feet (200 m) farther inland. So the Jogan event may have been even larger. The consensus about how enormous these waves could get was simply wrong: the most likely event was not the worst-case scenario.

"For people aware of paleo-tsunami work, it was not a huge surprise," said Lori Dengler. "But that information had definitely not been incorporated into the thinking of either the seismology community in Japan or the mainstream hazard folks. And it certainly had not been included in any of the planning for nuclear power plants."

Clearly some of the basic assumptions about what to expect from a subduction zone were wrong as well. Until Tohoku many seismologists thought old rocks like those along the Japan Trench were not likely to stick together for long periods of time. Only hot, fresh rocks in young subduction zones like Cascadia's fault were supposed to get stuck for hundreds of years and generate monster quakes. This is why computer models assumed the shallow part of the Japan Trench—where the fault comes to the surface of the sea floor—was not locked and therefore not likely to generate a monster shock. The realization that this idea is wrong has huge implications because the same assumption has been applied to many other subduction zones around the world.

"There are two big lessons from Tohoku for my nickel," Chris Goldfinger observed. "The first is that you can't base your estimates of hazard on short instrumental and historical records. Their giants come at thousand-year intervals, roughly, so even a thousand-year history was inadequate.

"The second lesson is that all previous seismological theories that relate simple things like plate age and subduction velocity to the maximum size of an earthquake . . . are now out the window. This is a very big deal, because these rather poorly justified models have been used to predict which systems globally are giant earthquake producers. And many regions were taken off the list on this basis. Tohoku was one of them. Now, all of them have to go back *on* the list and be reconsidered. Places like Java, Peru, New Zealand, et cetera. These are all places that haven't really worried about magnitude 9 earthquakes. Just as northeastern Japan didn't worry about them."

If any of this sounds familiar, recall that here in North America, scientists for years underestimated the potential for monster quakes from Cascadia's fault for exactly the same reasons. They had not looked far enough back into the geologic past for evidence of much bigger ruptures. But the evidence was always there. And in another parallel with Japan's tsunami history, it turns out that Cascadia's fault has also left tsunami footprints that were much larger than all the others in recent history.

Several years ago, when he first showed me the mud cores at his lab in Corvallis, Oregon, Goldfinger drew my attention to a sample from the great Cascadia earthquake of 1700. A magnitude 9 megathrust jolt had triggered an underwater mudslide along the continental slope, which dumped a significant swath of turbulent sand and gunk on the deep-sea floor. It was plain to see, in the sliced-open core sample, a darkish gray-brown layer nearly six inches thick.

Then Goldfinger pointed to another sample that was huge by comparison. "What's this?" he asked rhetorically, with a nervous laugh. Evidently an even larger quake and tsunami happened roughly 5,800 years ago, and it may have had *three times* as much energy as the magnitude 9 of 1700. Worse yet, there was another giant just like it roughly 8,800 years ago.

As monstrous as the 1700 Cascadia quake was—it wiped out native villages along the West Coast and sent deadly waves all the way to Japan—it was not the worst-case scenario for North America. It was only an "average-size" disaster. The much larger mega-tsunamis from Cascadia's prehistoric past have so far been ignored by the experts who estimate risk and plan for the future—a carbon copy of what just happened in Japan.

"[The Tohoku quake]at first seems like a rogue earthquake," said Goldfinger, "but in geology if you have something happen once, you immediately know that it has probably happened a thousand times. And you just didn't know about it."

Well, now we do know. The next obvious question is, what can we do about it? How well can any nation defend itself against a "thousand-year event"? Robert Geller, an American geologist at the University of Tokyo, has commented that "no conceivable economically realistic countermeasures could have precluded damages from a magnitude 9 earthquake and the resulting tsunami." But the civil engineers who inspected the damage agreed that well-built buildings saved lives. The seawalls may have failed, but even more people would have died if the barriers hadn't been there to slow the water down. Although devastation was widespread and the outcome tragic, strict construction standards did pay off.

Millions of people around the world saw huge high-rise towers in Tokyo swaying back and forth on live television—yet none of the towers collapsed. It was mostly smaller wood-frame houses that got swept away by the tsunami and destroyed. The EERI and ASCE reports noted that even in the worst-hit areas of northeastern Japan, the larger reinforced-concrete buildings that survived "did not appear to have significant structural damage from either the earthquake or tsunami . . . This provides some encouragement regarding the potential resilience of larger modern buildings with robust seismic designs and scour/uplift-resistant foundations."

The same was mostly true of the massive 2010 quake in Chile. Seismologist Garry Rogers of the Geological Survey of Canada described both recent events as "success posters for modern building codes." The point he stressed, however, is that seismic regulations are primarily designed to save human life, not to guarantee that a building will survive without a scratch. "Yes, there was damage to some modern buildings in Chile (and one collapse), but building codes are 'life safety' codes, not 'no damage' codes," Rogers explained. "They did the job they were supposed to do."

In the aftermath of the 2010 Chile quake, Lori Dengler and many other scientists felt a sense of optimism. Most people there escaped to higher ground as soon as the shaking stopped. They knew what to do and did not wait for anyone to tell them; they just did it.

"Chile *got* it," Dengler told me emphatically. Their story "was a success from almost any perspective. In Chile there were problems, but we all kind of felt like, yeah, they did a really good job. Engineering codes work. Education works."

But in the next breath she added, "That confidence that a lot of us had for Chile definitely evaporated in Japan. It's very sad . . . The one thing I'm absolutely sure of is that the next major event will surprise us in yet another way we haven't really thought about." Considering the parallels and implications for California, the Pacific Northwest, and the west coast of Canada, she added, "We need to be really humble in this business . . . That loss of confidence—it's significant . . . And it will happen to us."

But it seems Professor Dengler's spirit never stays down for long. She's the one who told me the wonderful story about the people of Langi Village on Simeulue Island who managed to escape the Indian Ocean tsunami of 2004 because their oral history included a strong memory of the last time the earth shook. The village elders knew exactly what to do. They moved every man, woman, and child to a prearranged sanctuary on higher ground. Not a single life was lost—even though they had only eight minutes before the first wave arrived.

Dengler writes an occasional column for the *Times-Standard*, a newspaper in Eureka on the northwest coast of California. After thinking about the risk that we who live in the shadow of Cascadia's fault must face, she came to pretty much the same conclusion I have. She wrote that "the North Coast is a special place and I feel extremely fortunate to live here. We have some of the highest volunteerism rates in the nation, we have an amazing entrepreneurial spirit, and we lead the nation in our tsunami preparedness efforts. It's not an accident or coincidence. It is because of our natural setting, not in spite of it. It creates self-reliance, and we should be thankful for that. If the Simeulue Islanders can do it, so can we."

As for the tragic death toll in Japan, recall that even though twenty thousand people died in a region with a population of millions, the vast *majority* of folks living there *did survive* the worst disaster in a thousand years. They're already well on their way toward cleaning up the debris and restarting their lives. Their experience has taught the whole world important new things about earthquake and tsunami science. The question now is, will we in North America listen and learn?

ACKNOWLEDGMENTS

In the course of filming a CBC documentary about Hungarians who fled to Canada after the 1956 uprising, I met Anna Porter, who escaped her Budapest home as a teenager and passed through New Zealand and the UK en route to becoming a successful author and publisher in Toronto. The room was lit and we were ready to shoot an introductory sequence of her working on final revisions to the manuscript of *Kastner's Train*. But when Anna searched her handbag, she discovered the manuscript had been left at home. In a pinch I found an acceptable substitute, a prop. I reached into my backpack and handed Anna the first draft of a film proposal about earthquakes and tsunamis along the Cascadia Subduction Zone, the documentary Bette and I hoped to produce the following summer.

Anna obligingly paged through the synopsis, marking it up with her red pen while the camera crew captured a serious book editor at work. Much to my good fortune, she wasn't just faking it. Anna was reading and paying attention. Finally she mumbled under her breath, "You realize there's a book in this, don't you?" And so here we are. Thanks ever so much to Anna Porter, not only for planting the seed but also for personally talking about it to her friends and colleagues in the publishing industry. I'm convinced she helped get this book over the transom and near the top of the slush pile in record time.

More than two years later, when I had written a draft of most of the chapters, I returned to the source, to the quake hunters, oceanographers, and other seismic sleuths who did the muddy detective work that unraveled the mysteries of Cascadia's fault. A dozen of them found time in their hectic schedules to read and comment on parts of the manuscript. I want to acknowledge and sincerely thank John Adams, Brian Atwater, Eddie Bernard, Gary Carver, Lori Dengler, Chris Goldfinger, George Plafker, Garry Rogers, Mike Schmidt, Vasily Titov, Kelin Wang, and Bob Yeats for their insights, clarifications, and valuable suggestions. In the final draft, of course, all errors of fact or interpretation are my responsibility.

I'd also like to express my deep thanks and appreciation to Harper-Collins and editor Jim Gifford for making this journey and the learning curve such a pleasant and rewarding experience. Jim brought fresh eyes and an organized mind to a dense thicket of complex material. The book is all the better for his efforts.

And finally, I thank Bette, my partner in life and everything else. Without her encouragement and support, without her voluminous research—she can find anything!—and razor sharp attention to detail, this book would never have been completed. She carried the heaviest burden of several other projects we were committed to (not the least of which was organizing and running our lives) so that I could indulge in the luxury of chasing a story down convoluted alleys to its logical end. BT, I'm eternally happy to be on your team.

SUGGESTIONS FOR FURTHER READING

A chronology of scientific articles, proceedings, working papers, and books that trace the evolution and mystery of Cascadia's fault from early continental drift to plate tectonics and to the current reality.

1924

Wegener, Alfred. *The Origin of Continents and Oceans.* New York: E. P. Dutton, 1924.

1949

Benioff, Hugo. "Seismic Evidence for the Fault Origin of Oceanic Deeps." *Bulletin of the Geological Society of America* 60 (December 1949): 1837–66.

1962

Hess, H. H. "History of Ocean Basins." In *Petrologic Studies: A Volume to Honor A. F. Buddington*, edited by A. E. J. Engel, Harold L. James, and B. F. Leonard, 599–620. Princeton: Princeton University Press, 1962.

1963

Vine, F. J., and D. H. Matthews. "Magnetic Anomalies over Ocean Ridges." *Nature* 199, no. 4897 (September 1963): 947–49.

Wilson, J. Tuzo. "Are the Continents Drifting? A New Look at a Controversial Question." *The UNESCO Courier* no. 10 (October 1963): 3–11.

———. "Hypothesis of Earth's Behaviour." *Nature* 198, no. 4884 (June 1963): 925–29.

1964

Benioff, Hugo. "Earthquake Source Mechanisms: Although Progress Has Been Made in the Understanding of Earthquakes, Many Problems Remain." *Science* 143, no. 3613 (March 1964): 1399–1406.

1965

Plafker, George. "Tectonic Deformation Associated with the 1964 Alaska Earthquake." *Science* 148, no. 3678 (June 1965): 1675–87.

Press, Frank, and David Jackson. "Alaskan Earthquake, 27 March 1964: Vertical Extent of Faulting and Elastic Strain Energy Release." *Science* n.s. 147, no. 3660 (February 1965): 867–68.

Wilson, J. Tuzo. "A New Class of Faults and Their Bearing on Continental Drift." *Nature* 207, no. 4995 (July 1965): 343–47.

———. "Transform Faults, Oceanic Ridges and Magnetic Anomalies Southwest of Vancouver Island." *Science* 150, no. 3695 (October 1965): 482–85.

1968

Isacks, Bryan, and Jack Oliver. "Seismology and the New Global Tectonics." *Journal of Geophysical Research* 73, no. 18 (September 1968): 5855–99.

1970

Griggs, G. B., and L. D. Kulm. "Sedimentation in Cascadia Deep-Sea Channel." *Geological Society of America Bulletin* 81 (May 1970): 1361–84.

Plafker, George, and J. C. Savage. "Mechanism of the Chilean Earthquakes of May 21 and 22, 1960." *Geological Society of America Bulletin* 81 (April 1970): 1001–30.

1972

Committee on the Alaska Earthquake. *The Great Alaska Earthquake of 1964: Seismology and Geodesy.* Washington, DC: National Academy of Sciences, 1972.

1976

Riddihough, R. P., and R. D. Hyndman. "Canada's Active Western Margin: The Case for Subduction." *Geoscience Canada* 3, no. 4 (1976): 269–78.

1979

Ando, Masataka, and Emery I. Balazs. "Geodetic Evidence for Aseismic Subduction of the Juan de Fuca Plate." *Journal of Geophysical Research* 84, no. B6 (June 1979): 3023–27.

1981

Savage, J. C., M. Lisowski, and W. H. Prescott. "Geodetic Strain Measurements in Washington." *Journal of Geophysical Research* 86, no. B6 (June 1981): 4929–40.

1982

Reilinger, Robert, and John Adams. "Geodetic Evidence for Active Landward Tilting on the Oregon and Washington Coastal Ranges." *Geophysical Research Letters* 9, no. 4 (April 1982): 401–3.

1984

Adams, John. "Active Deformation of the Pacific Northwest Continental Margin." *Tectonics* 3, no. 4 (August 1984): 449–72.

Heaton, Thomas H., and Hiroo Kanamori. "Seismic Potential Associated with Subduction in the Northwestern United States." *Bulletin of the Seismological Society of America* 74, no. 3 (June 1984): 933–41.

1985

Bakun, W. H., and A. G. Lindh. "The Parkfield, California, Earthquake Prediction Experiment." *Science* 229, no. 4714 (August 1985): 619–24.

Heaton, Thomas H., and Parke D. Snavely Jr. "Possible Tsunami along the Northwestern Coast of the United States Inferred from Indian Traditions." *Bulletin of the Seismological Society of America* 75, no. 5 (October 1985): 1455–60.

1986

Heaton, Thomas H., and Stephen H. Hartzell. "Source Characteristics of Hypothetical Subduction Earthquakes in the Northwestern United States." *Bulletin of the Seismological Society of America* 76, no. 3 (June 1986): 675–708.

Mitchell, Denis, John Adams, Ronald H. DeVall, Robert C. Lo, and Dieter Weichert.

"Lessons from the 1985 Mexican Earthquake." *Canadian Journal of Civil Engineering* 13 (1986): 535–57.

1987

Atwater, Brian F. "Evidence for Great Holocene Earthquakes along the Outer Coast of Washington State." *Science* 236, no. 4804 (May 1987): 942–44.

Carver, G. A., and R. M. Burke. "Late Pleistocene and Holocene Paleoseismicity of Little Salmon and Mad River Thrust Systems, NW California: Implications to the Seismic Potential of the Cascadia Subduction Zone." *Geological Society of America Bulletin, Abstracts with Programs*, no. 125029 (1987).

Heaton, Thomas H., and Stephen H. Hartzell. "Earthquake Hazards on the Cascadia Subduction Zone." *Science* 236, no. 4798 (April 1987): 162–68.

Yeats, Robert S. "Summary of Symposium on Oregon's Earthquake Potential Held February 28, 1987, at Western Oregon State College in Monmouth." *Oregon Geology* 49, no. 8 (August 1987): 97–98.

1988

Dragert, H., and G. C. Rogers. "Could a Megathrust Earthquake Strike Southwestern British Columbia?" *Geos* 3 (1988): 5–8.

Nance, John J. *On Shaky Ground: An Invitation to Disaster.* New York: William Morrow, 1988.

Rogers, Garry C. "An Assessment of the Megathrust Earthquake Potential of the Cascadia Subduction Zone." *Canadian Journal of Earth Sciences* 25 (1988): 844–52.

1990

Adams, John. "Paleoseismicity of the Cascadia Subduction Zone: Evidence from Turbidites off the Oregon–Washington Margin." *Tectonics* 9, no. 4 (August 1990): 569–83.

Darienzo, Mark E., and Curt D. Peterson. "Episodic Tectonic Subsidence of Late Holocene Salt Marshes, Northern Oregon Central Cascadia Margin." *Tectonics* 9, no. 1 (1990): 1–22.

Dragert, H., and M. Lisowski. "Crustal Deformation Measurements on Vancouver Island, British Columbia: 1976 to 1988." In *Proceedings of a Symposium Held within the General Meeting of the International Association of Geodesy in Edinburgh, Scotland, August 3–5, 1989*, edited by Ivan I. Mueller, 241–50. New York: Springer-Verlag, 1990.

1991

Atwater, Brian F., and David K. Yamaguchi. "Sudden, Probably Coseismic Submergence of Holocene Trees and Grass in Coastal Washington State." *Geology* 19 (July 1991): 706–9.

1992

Goldfinger, C., L. D. Kulm, R. S. Yeats, B. Applegate, M. MacKay, and G. F. Moore. "Transverse Structural Trends along the Oregon Convergent Margin: Implications for Cascadia Earthquake Potential." *Geology* 20 (February 1992): 141–44.

1993

Hyndman, R. D., and K. Wang. "Thermal Constraints on the Zone of Major Thrust Earthquake Failure: The Cascadia Subduction Zone." *Journal of Geophysical Research* 98, no. B2 (1993): 2039–60.

Oppenheimer, D., G. Beroza, G. Carver, L. Dengler, J. Eaton, L. Gee, F. Gonzalez et al. "The Cape Mendocino, California, Earthquakes of April 1992: Subduction at the Triple Junction." *Science* 261, no. 5120 (July 1993): 433–38.

1994

Clague, John J., and Peter T. Bobrowsky. "Evidence for a Large Earthquake and Tsunami 100–400 Years ago on Western Vancouver Island, British Columbia." *Quaternary Research* 41 (1994): 176–84.

Darienzo, M. E., C. D. Peterson, and C. Clough. "Stratigraphic Evidence for Great Subduction-Zone Earthquakes at Four Estuaries in Northern Oregon, U.S.A." *Journal of Coastal Research* 10, no. 4 (1994): 850–76.

Dragert, H., R. D. Hyndman, G. C. Rogers, and K. Wang. "Current Deformation and the Width of the Seismogenic Zone of the Northern Cascadia Subduction Thrust." *Journal of Geophysical Research* 99 (1994): 653–658.

1995

Hyndman, R. D., and K. Wang. "The Rupture Zone of Cascadia Great Earthquakes from Current Deformation and the Thermal Regime." *Journal of Geophysical Research* 100, no. B11 (November 1995): 22133–54.

McCaffrey, Robert, and Chris Goldfinger. "Forearc Deformation and Great Subduction Earthquakes: Implications for Cascadia Offshore Earthquake Potential." *Science* 267, no. 5199 (February 1995): 856–59.

Nelson, Alan R., Brian F. Atwater, Peter T. Bobrowsky, Lee-Ann Bradley, John J. Clague, Mark E. Darienzo, Wendy C. Grant et al. "Radiocarbon Evidence for Extensive Plate-Boundary Rupture about 300 Years ago at the Cascadia Subduction Zone." *Nature* 378 (November 1995): 371–74.

1996

Adams, John. "Great Earthquakes Recorded by Turbidites off the Oregon–Washington Coast." In *Assessing Earthquake Hazards and Reducing Risk in the Pacific Northwest,* vol. 1, edited by Albert M. Rogers, Timothy J. Walsh, William J. Knockelman, and George R. Priest, 147–58. U.S. Geological Survey Professional Paper 1560. Washington, DC: U.S. Government Printing Office, 1996.

Ikeda, Yasutaka. "Evaluating Seismic-Risk Potential of Active Faults: What Should We Do Now?" *Active Fault Research* 15 (1996): 93–99.

Minor, Rick, and Wendy C. Grant. "Earthquake-Induced Subsidence and Burial of Late Holocene Archaeological Sites, Northern Oregon Coast." *American Antiquity* 61 (1996): 772–81.

Satake, Kenji, Kunihiko Shimazaki, Yoshinobu Tsuji, and Kazue Ueda. "Time and Size of a Giant Earthquake in Cascadia Inferred from Japanese Tsunami Records of January 1700." *Nature* 379 (January 1996): 246–49.

1997

Atwater, B. F., and E. Hemphill-Haley. *Recurrence Intervals for Great Earthquakes of the Past 3500 Years at Northeastern Willapa Bay, Washington.* U.S. Geological Survey Professional Paper 1576. Washington, DC: U.S. Government Printing Office, 1997.

Geller, Robert J., David D. Jackson, Yan Y. Kagan, and Francesco Mulargia. "Earthquakes Cannot Be Predicted." *Science* 275, no. 5306 (March 1997): 1616–17.

Goldfinger, C., L. D. Kulm, R. S. Yeats, L. McNeill, and C. Hummon. "Oblique Strike-Slip Faulting of the Central Cascadia Submarine Forearc." *Journal of Geophysical Research* 102, no. B4 (1997): 8217–43.

Yamaguchi, David K., Brian F. Atwater, Daniel E. Bunker, Boyd E. Benson, and Marion

S. Reid. "Tree-Ring Dating the 1700 Cascadia Earthquake." *Nature* 389 (October 1997): 922.

1998

Carver, Deborah H. *Native Stories of Earthquake and Tsunamis, Redwood National Park, California.* Crescent City, CA: U.S. National Park Service, 1998.

1999

Goldfinger, C., and C. H. Nelson. "Holocene Recurrence of Cascadia Great Earthquakes Based on the Turbidite Event Record." *Eos, Transactions, American Geophysical Union* 80 (1999): 1024.

Oreskes, Naomi. *The Rejection of Continental Drift: Theory and Method in American Earth Science.* New York: Oxford University Press, 1999.

2000

Nelson, C. H., C. Goldfinger, and J. E. Johnson. "Turbidite Event Stratigraphy and Implications for Cascadia Basin Paleoseismicity." In *Penrose Conference 2000 Great Cascadia Earthquake Tricentennial,* edited by J. J. Clague, B. F. Atwater, K. Wang, Y. Wang, and I. Wong, 156. Special Paper 33. Portland: Oregon Department of Geology and Mineral Industries, 2000.

Nishimura, Takuya, Satoshi Miura, Kenji Tachibana, Keiichi Hashimoto, Toshiya Sato, Syuitchiro Hori, Eiju Murakami et al. "Distribution of Seismic Coupling on the Subducting Plate Boundary in Northeastern Japan Inferred from GPS Observations." *Tectonophysics* 323 (2000): 217–238.

2001

Dragert, Herb, Kelin Wang, and Thomas S. James. "A Silent Slip Event on the Deeper Cascadia Subduction Interface." *Science* 292, no. 5521 (May 2001): 1525–28.

Goldfinger, C., C. H. Nelson, and J. E. Johnson. "Temporal Patterns of Turbidites Offshore the Northern San Andreas Fault and Correlation to Paleoseismic Events Onshore." *Eos, Transactions, American Geophysical Union* 82 (2001): F934.

Minoura, K., F. Imamura, D. Sugawara, Y. Kono, and T. Iwashita. "The 869 Jogan Tsunami Deposit and Recurrence Interval of Large-Scale Tsunami on the Pacific Coast of Northeast Japan." *Journal of Natural Disaster Science* 23, no. 2 (2001): 83–88.

2002

Garrison-Laney, C. E., H. F. Abramson, and G. A. Carver. "Late Holocene Tsunamis Near the Southern End of the Cascadia Subduction Zone." *Seismological Research Letters* 73, no. 2 (2002): 248.

McMillan, Alan D., and Ian Hutchison. "When the Mountain Dwarfs Danced: Aboriginal Traditions of Paleoseismic Events along the Cascadia Subduction Zone of Western North America." *Ethnohistory* 49, no. 1 (2002): 42–68.

2003

Goldfinger, Chris, C. Hans Nelson, Joel E. Johnson, and the Shipboard Scientific Party. "Deep-Water Turbidites as Holocene Earthquake Proxies: The Cascadia Subduction Zone and Northern San Andreas Fault System." *Annals of Geophysics* 46, no. 5 (October 2003): 1169–94.

Oreskes, Naomi, ed., with Homer Le Grand. *Plate Tectonics: An Insider's History of the Modern Theory of the Earth.* Boulder, CO: Westview Press, 2003.

Rogers, Garry, and Herb Dragert. "Episodic Tremor and Slip on the Cascadia Subduction Zone: The Chatter of Silent Slip." *Science* 300, no. 5627 (June 2003): 1942–43.

Satake, Kenji, Kelin Wang, and Brian F. Atwater. "Fault Slip and Seismic Moment of the 1700 Cascadia Earthquake Inferred from Japanese Tsunami Descriptions." *Journal of Geophysical Research* 108, no. B11 (2003): 2535.

Winchester, Simon. *Krakatoa: The Day the World Exploded; August 27, 1883.* New York: Harper Perennial, 2003.

2004

Dragert, H., K. Wang, and G. Rogers. "Geodetic and Seismic Signatures of Episodic Tremor and Slip in the Northern Cascadia Subduction Zone." *Earth Planets Space* 56 (2004): 1143–50.

Yeats, Robert S. *Living with Earthquakes in the Pacific Northwest: A Survivor's Guide.* 2nd ed. Corvallis: Oregon State University Press, 2004.

2005

Atwater, Brian F., Musumi-Rokkaku Satoko, Satake Kenji, Tsuji Yoshinobu, Ueda Kazue, and David K. Yamaguchi. *The Orphan Tsunami of 1700: Japanese Clues to a Parent Earthquake in North America.* Reston, VA: U.S. Geological Survey in association with University of Washington Press, 2005.

Lindh, Allan Goddard. "Opinion: Success and Failure at Parkfield." *Seismological Research Letters* 76, no. 1 (January–February 2005): 3–6.

Ludwin, Ruth S., Robert Dennis, Deborah Carver, Alan D. McMillan, Robert Losey, John Clague, Chris Jonientz-Trisler et al. "Dating the 1700 Cascadia Earthquake: Great Coastal Earthquakes in Native Stories." *Seismological Research Letters* 76, no. 2 (March 2005): 140–48.

Winchester, Simon. *A Crack in the Edge of the World: America and the Great California Earthquake of 1906.* New York: Harper Perennial, 2005.

2006

Clague, John, Chris Yorath, Richard Franklin, and Bob Turner. *At Risk: Earthquakes and Tsunamis on the West Coast.* Vancouver: Tricouni Press, 2006.

Goldfinger, C., and L. C. McNeill. "Sumatra and Cascadia: Parallels Explored." *Eos, Transactions, American Geophysical Union,* Fall Meeting Supplement, Abstract U53A-0026, v. 87 (2006).

Wang, Kelin, Qi-Fu Chen, Shihong Sun, and Andong Wang. "Predicting the 1975 Haicheng Earthquake." *Bulletin of the Seismological Society of America* 96, no. 3 (June 2006): 757–95.

2007

Bernard, E. N., L. A. Dengler, and S. C. Yim. *National Tsunami Research Plan: Report of a Workshop Sponsored by NSF/NOAA.* NOAA Technical Memorandum OAR PMEL-133, March. Seattle: U.S. Department of Commerce, 2007.

Goldfinger, C., Yusuf S. Djajadihardja, and the Shipboard Scientific Research Party. "Cruise Report, PaleoQuakes07, 7 May–14 June, 2007." Paleoseismologic Studies of the Sunda Subduction Zone. Oregon State University Active Tectonics Laboratory and Agency for the Assessment and Application of Technology, Indonesia, 2007.

Satake, Kenji, and Brian F. Atwater. "Long-Term Perspectives on Giant Earthquakes and Tsunamis at Subduction Zones." *Annual Review of Earth and Planetary Sciences* 35 (2007): 349–75.

2008

Field, Edward H., Kevin R. Milner, and the 2007 Working Group on California Earthquake Probabilities. *Forecasting California Earthquakes: What Can We Expect in the Next 30 Years?* USGS Fact Sheet 2008–3027. http://pubs.usgs.gov/fs/2008/3027.

2010

Goldfinger, C., C. H. Nelson, A. Morey, J. E. Johnson, J. Gutierrez-Pastor, A. T. Eriksson, E. Karabanov et al. *Turbidite Event History: Methods and Implications for Holocene Paleoseismicity of the Cascadia Subduction Zone.* U.S. Geological Survey Professional Paper 1661-F. Reston, VA: U.S. Geological Survey, 2010.

2011

Avouac, Jean-Philippe. "The Lessons of Tohoku-Oki." *Nature* 475 (July 2011): 300–301.

Cyranoski, David. "Japan Faces Up to Failure of Its Earthquake Preparations." *Nature* 471 (March 2011): 556–557.

Dengler, Lori, Megumi Sugimoto, Rick Wilson, and Bruce Jafe. "The Japan Tohoku Tsunami of March 11, 2011." *EERI Special Earthquake Report* (November 2011). http://www.eqclearinghouse.org/2011-03-11-sendai/files/2011/11/Japan-eq-report-tsunami2.pdf.

Goto, Kazuhisa, Catherine Chagué-Goff, Shigehiro Fujino, James Goff, Bruce Jaffe, Yuichi Nishimura, Bruce Richmond et al. "New Insights of Tsunami Hazard from the 2011 Tohoku-Oki Event." *Marine Geology* 290, no. 1–4 (2011): 46–50.

Ozawa, Shinzaburo, Takuya Nishimura, Hisashi Suito, Tomokazu Kobayashi, Mikio Tobita, and Tetsuro Imakiire. "Coseismic and Postseismic Slip of the 2011 Magnitude-9 Tohoku-Oki Earthquake." *Nature* 475 (July 2011): 373–376.

Robertson, Ian, and Gary Chock. "The Tohoku, Japan, Tsunami of March 11, 2011: Effects on Structures." *EERI Special Earthquake Report* (September 2011). http://www.eqclearinghouse.org/2011-03-11-sendai/files/2011/03/Japan-Tohoku-report-tsunami-bldgs.pdf.

Sagiya, Takeshi, Hiroo Kanamori, Yuji Yagi, Masumi Yamada, and Jim Mori. "Rebuilding Seismology." *Nature* 473 (May 2011): 146–148.

Sato, Mariko, Tadashi Ishikawa, Naoto Ujihara, Shigeru Yoshida, Masayuki Fujita, Masashi Mochizuki, and Akira Asada. "Displacement above the Hypocenter of the 2011 Tohoku-Oki Earthquake." *Science* 332, no. 6036 (June 2011): 1395.

INDEX

NOAA pays attention to CSZ, 167–68

PG&E nuclear reactor, 65

prediction debate, 264

raising public awareness, 285, 314–17

Simeulue Island story, 315–17, 335–36

Tilly Smith story, 315

Tohoku-Oki earthquake and tsunami of 2011, 325, 326, 327, 328, 332, 335–36

Dennis, Robert, 197–98

detection equipment

bottom pressure recorder (BPR), 186

creepmeters, 25, 256, 258

seismographs, 14, 38, 75, 185, 186, 219, 236, 256, 258

seismometers, 44, 188, 255, 259, 261, 323

tiltmeters, 258

tsunameters, 186–87, 188

DeVall, Ron, 15–17

documentaries

Quake Hunters, 196

ShockWave, 61, 116, 231, 289, 313

Douglas, Robb, 3–4, 6–7, 15

Dragert, Herb

crustal compression, measuring, 115, 117–18, 124, 129, 156–58, 178, 213, 214

earthquake history of CSZ, 157

earthquake triggering, 262

episodic tremor and slip (ETS), 215–19, 258, 260, 262

locked subduction zone debate, 157, 218

locked tectonic plates debate, 118

Plate Boundary Observatory project, 261

silent slip, 215, 218

Wilson, J. Tuzo, 115, 118, 157

E

Earthquake Engineering Research Institute (EERI), 329, 334

Earthquake Engineering Research Laboratory, 296–97

Earthquake Prediction and Public Policy, 235 (*see also* prediction debate)

Earthquake Research Institute (Japan), 253

earthquake triggering, *xiii, xix,* 231–32, 262, 263

EarthScope, 259–60

Plate Boundary Observatory project, 259–60, 261

SAFOD project, 258–59

USArray, 259

van der Vink, Greg, 260

East Pacific Rise, 59

Eel River, 166

elastic rebound hypothesis, 249

"Elvis" fault, 190, 225

epicenters, 27, 38, 45, 51, 163, 165, 178, 179, 181, 185, 243, 255, 315–16, 322, 323, 326

episodic tremor and slip (ETS)

Adams, John, 262

Dragert, Herb, 215–19, 258, 260, 262

Goldfinger, Chris, 261–62, 263

Rogers, Garry, 219–20, 258, 260, 261, 262–63

silent slip, 215, 217, 218

Wang, Kelin, 217–18

Eureka, *xix,* 61, 62, 63, 64, 66, 336

new faults discovered, 73, 77, 143

nuclear power plant, 65, 66, 68, 77

Ewa Beach, 184